A Gift of Wings

The joy of flight.
The magic of flight.
The meaning of flight.
The endless challenge and infinite rewards of flight.
This is what Richard Bach writes about.

For all who wish to rise above their earth-bound existences to feast on the freedom and adventure that Richard Bach knows and loves and recreates so magnificently, this book offers—

A Gift of Wings

DEC - - 1998

Editor's note

When I wrote Richard Bach the letter that resulted in the publication of *Jonathan Livingston Seagull,* I knew him very well, although I had never met him in person or spoken to him or written him before. I had read his first novel, *Stranger to the Ground,* and those 173 pages with him in a jet fighter plane over Europe told me enough to make me write, more than six years later, "I have a very special feeling that you could do a work of fiction that would somehow speak for the next few decades. . . ."

There is a lot about flying in this book, but much more about Richard Bach and his last fifteen years of seeking answers and finding some. For anyone who cares to know who he is, it is all here. The reminiscences and stories were arranged by the author for pace and enjoyment in reading; they are not in chronological order. For the reader who wants to place this life in sequence, the last pages of the book record the year each story was written.

E.F.

A Gift of Wings

Richard Bach

A Dell/Eleanor Friede Book

921
BACH

A DELL/ELEANOR FRIEDE BOOK
Published by
Dell Publishing
a division of
Bantam Doubleday Dell Publishing Group, Inc.
1540 Broadway
New York, N.Y. 10036

If you purchased this book without a cover you should be aware that this book is stolen property. It was reported as "unsold and destroyed" to the publisher and neither the author nor the publisher has received any payment for this "stripped book."

Portions of this book appeared previously in *Flying* (Ziff-Davis Publishing Company), *Air Progress* (Slawson Communications, Inc.), *Private Pilot* (Peterson Publishing Company), *Argosy* (Popular Publications, Inc.), *Sport Flying*, and *Air Facts*.

"Across the country on an oil pressure gage" originally appeared as "Westward the—What kind of airplane is that anyway?" Copyright © 1964 by Ziff-Davis Publishing Company. "Think black" Copyright © 1962 by Ziff-Davis Publishing Company. Both reprinted by permission of *Flying* magazine and the Ziff-Davis Publishing Company.

Copyright © 1974 by Alternate Futures Inc., PSP

Excerpts from WIND, SAND AND STARS and THE LITTLE PRINCE by Antoine de Saint-Exupéry are used by permission of Harcourt Brace Jovanovich, Inc.

All rights reserved. No part of this book may be reproduced in any form or by any means without the prior written permission of Delacorte Press, New York, New York, excepting brief quotes used in connection with reviews written specifically for inclusion in a magazine or newspaper.

The trademark Dell® is registered in the U.S. Patent and Trademark Office.

ISBN: 0-440-20432-1

November 1975

20

OPM

Contents

It is said that we have ten seconds

when we wake of a morning, to remember what it was we dreamed the night before. Notes in the dark, eyes closed, catch bits and shards and find what the dreamer is living, and what the dreaming self would say to the self awake.

I tried that for a while with a tape recorder, talking my dreams into a little battery-powered thing by the pillow, the moment I woke. It didn't work. I remembered for a few seconds what had happened in the night, but I could never understand later what the sounds on the tape were saying. There was only this odd croaking tomb voice, hollow and old as some crypt door, as though sleep were death itself.

A pen with paper worked better, and when I learned not to write one line on top of another, I began to know about the travels of that part of me that never sleeps at all. Lots of mountains, in dream country, lots of flying going on, lots of schools, lots of oceans plowing into high cliffs, lots of strange trivia and now and then a rare moment that might have been from a life gone by, or from one yet to be.

It wasn't much later that I noticed that my days were dreams themselves, and just as deeply forgotten. When I couldn't remember what happened last Wednesday, or even last Saturday, I began keeping a journal of days as well as of nights, and for a long time I was afraid that I had forgotten most of my life.

When I gathered up a few cardboard boxes of writing, though, and put together my favorite best stories of the last fifteen years into this book, I found that I hadn't forgotten quite so much, after all. Whatever sad times bright times strange fantasies struck me as I flew, I had written—stories and articles instead of pages in a journal, several hundred of them in all. I had promised when I bought my first typewriter that I would never write about anything that didn't matter to me, that didn't make some difference in my life, and I've come pleasantly close to keeping that promise.

There are times in these pages, however, that are not very well written—I have to throw my pen across the room to keep from rewriting *There's Something the Matter with Seagulls* and *I've Never Heard the Wind,* the first stories of mine to sell to any magazine. The early stories are here because something that mattered to the beginner can be seen even through the awkward writing, and in the ideas he reached for are some learning and perhaps a smile for the poor guy.

Early in the year that my Ford was repossessed, I wrote a note to me across some calendar squares where a distant-future Richard Bach might find it:

> *How did you survive to this day? From here it looks like a miracle was needed. Did the Jonathan Seagull book get published? Any films?*
> *What totally unconceived new projects? Is it all better and happier? What do you think of my fears?*
>
> —RB *22 March 1968*

Maybe it's not too late to appear in a smoke puff and answer his questions.

> *You survived because you decided against quitting when the battle wasn't much fun . . . that was the only miracle required. Yes, Jonathan finally was published. The film ideas, and a few others you hadn't thought of, are just beginning. Please don't waste your time worrying or being afraid.*

Angels are always saying that sort of thing: don't fret, fear not, everything's going to be OK. Me-then would probably have frowned at me-now and said, "Easy words for you, but I'm running out of food and I've been broke since Tuesday!"

Maybe not, though. He was a hopeful and trusting person. Up to a point. If I tell him to change words and paragraphs, cut this and add that, he'll ask that I get lost, please, just run along back into the future, that he knows very well how to say what he wants to say.

An old maxim says that a professional writer is an amateur who didn't quit. Somehow, maybe because he couldn't keep any other job for long, the awkward beginner became an unquitting amateur, and still is. I never could think of myself as a Writer, as a complicated soul who lives only for words in ink. In fact, the only time I can write is when some idea is so scarlet-fierce that it grabs me by the neck and drags me thrashing and screaming to the typewriter. I leave heel marks on the floors and fingernail scratches in the walls every inch of the way.

It took far too long to finish some of these stories. Three years to write *Letter from a God-fearing Man,* for instance. I'd hit that thing over and over, knowing it had to be written somehow, knowing there was a lot that mattered, that needed saying there. Forced to the typewriter, all I'd do was surround myself with heaps of crumpled paper, the way writers do in movies. I'd get up gnashing and snarling and go wrap myself around a pillow on the bed to try it longhand in a fresh notebook, a trick that sometimes works on hard stories. But the religion-of-flight idea kept coming out of my pencil the color of lead and ten times heavier and I'd mutter harsh words and crunch it up as though solemn bad writing can be crunched and thrown at a wall as easily as notebook paper.

But then one day there it was. It was the guys at the soap factory that made it work—without the crew at Vat Three who showed up out of nowhere, the story would be a wrinkled ball at some baseboard yet.

It took time to learn that the hard thing about writing is to let the story write itself, while one sits at the typewriter and does as little thinking as possible. It happened over and again, and the beginner learned—when you start puzzling over an idea, and slowing down on the keys, the writing gets worse and worse.

Adrift at Kennedy Airport comes to mind. The closest I steered to insanity was in that one story, originally planned as a book. As with *Letter,* the words kept swinging back to invisible dank boredom; all sorts of numbers and statistics kept appearing in the lines. It went on that way for nearly a year, days and weeks at the monster circus-airport, watching all the acts, satchels filling with popcorn research, pads of cotton-candy notes, and it all turned into gray chaff on paper.

When I decided at last that I didn't care what the book publisher wanted and that I didn't care what I wanted and that I was just going to go ahead and be naive and foolish and forget everything and *write,* that is when the story opened its eyes and started running around.

The book was rejected when the editor saw it charging across the playground without a single statistic on its back, but *Air Progress* printed it at once, as it was—not a book, not an article, not an essay. I don't know whether I won or lost that round.

Anyone who would print his loves and fears and learnings on the pages of magazines says farewell to the secrets of his mind and gives them to the world. When I wrote *The Pleasure of Their Company,* one side of this farewell was simple and clear: "The way to know any writer is not to meet him in person, but to read what he writes." The story put itself on paper out of a sudden realization . . . some of my closest friends are people I'll never meet.

The other side of this farewell to secrets took some years to see. What can you say to a reader who walks up at an airport knowing you better than he knows his own brother? It was hard to believe that I hadn't been confiding my inner life to a solitary typewriter, or even to a sheet of paper, but to living people who will occasionally appear and say hello. This is not all fun for one who likes lonely things like sky and aluminum and places that are quiet in the night. "HI THERE!" in what has always been a silent unseen place is a scary thing, no matter how well meant it's said.

I'm glad now that it was too late for me to call Nevil Shute on the telephone, or Antoine de Saint-Exupéry, or Bert Stiles, when I found that I loved who they are. I could only have frightened them with my praise, forced them to build glad-you-liked-the-book walls against my intrusions. I know them better, now, for never having spoken with them or never having met them at bookstore autograph parties. I didn't know this when *The Pleasure of Their Company* was written, but that's not a bad thing . . . new truths fit old ones without seams or squeaks.

Most of the stories here were printed in special-interest magazines. A few thousand people might have read them and thrown them away, or dropped them off in stacks at a Boy Scout paper drive. It's a quick world, magazine writing. Life there has the span of a May-fly's, and death is having no stories in print at all.

The best of my paper children are here, rescued from beneath tons of trash, saved from flame and smoke, alive again, leaping from castle walls because they believe that flying is a happy thing to do. I read them today and hear myself in an empty room: "There is a lovely story, Richard!" "Now *that* is what I call beautiful writing!" These make me laugh, and sometimes in some places they make me cry, and I like them for doing that.

Perhaps one or two of my children might be yours, too, and take your hand and maybe help you touch the part of your home that is the sky.

—RICHARD BACH
August 1973

People who fly

For nine hundred miles, I listened to the man in the seat next to mine on Flight 224 from San Francisco to Denver. "How did I come to be a salesman?" he said. "Well, I joined the Navy when I was seventeen, in the middle of the war . . ." And he had gone to sea and he was in the invasion of Iwo Jima, taking troops and supplies up to the beach in a landing craft, under enemy fire. Incidents many, and details of the time, back in the days when this man had been alive.

Then in five seconds he filled me in on the twenty-three years that came after the war: ". . . so I got this job with the company in 1945 and I've been here ever since."

We landed at Denver Stapleton and the flight was over. I said goodbye to the salesman, and we went our ways into the crowd at the terminal and of course I never saw him again. But I didn't forget him.

He had said it in so many words—the only real life he had known, the only real friends and real adventures, the only things worth remembering and reliving since he was born were a few scattered hours at sea in the middle of a world war.

In the days that led away from Denver, I flew light airplanes into little summer fly-ins of sport pilots around the country, and I thought of the salesman often and I asked myself time and again, what do *I* remember? What times of real friends and real adventure and real life would I go back to and live over again?

I listened more carefully than ever to the people around

me. I listened as I sat with pilots, now and then, clustered on the night grass under the wings of a hundred different airplanes. I listened as I stood with them in the sun and while we walked aimlessly, just for the sake of talking, down rows of bright-painted antiques and home-builts and sport planes on display.

"I suspect the thing that makes us fly, whatever it is, is the same thing that draws the sailor out to the sea," I heard. "Some people will never understand why and we can't explain it to them. If they're willing and have an open heart we can show them, but *tell* them we can't."

It's true. Ask "Why fly?" and I should tell you nothing. Instead, I should take you out to the grounds of an airport on a Saturday morning in the end of August. There is sun and a cloud in the sky, now, and here's a cool breeze hushing around the precision sculptures of lightplanes all washed in rainbows and set carefully on the grass. Here's a smell of clean metal and fabric in the air, and the swishing chug of a small engine spinning a little windmill of a propeller, making ready to fly.

Come along for a moment and look at a few of the people who choose to own and fly these machines, and see what kind of people they are and why they fly and whether, because of it, they might be a little bit different than anyone else in all the world.

I give you an Air Force pilot, buffing the silver cowl of a lightplane that he flies in his off-duty hours, when his eight-engine jet bomber is silent.

"I guess I'm a lover of flying, and above all of that tremendous rapport between a man and an airplane. Not just any man—let me exclude and be romantic—but a man who feels flight as his life, who knows the sky not as work or diversion, but as *home*."

Listen to a couple of pilots as one casts a critical eye on his wife in her own plane, practicing landings on the grass runway: "Sometimes I watch her when she thinks I'm gone. She kisses that plane on the spinner, before she locks the hangar at night."

An airline captain, touching up the wing of his homebuilt racer with a miniature paint bottle and a tiny brush. "Why fly? Simple. I'm not happy unless there's some air between me and the ground."

In an hour, we talk with a young lady who only this morning learned that an old two-winger has been lost in a hangar fire: "I don't think you're ever the same after seeing the world framed by the wings of a biplane. If someone had told me a year ago that I could cry over an airplane, I would have laughed. But I had grown to love that old thing . . ."

Do you notice that when these people talk about why they fly and the way that they think about airplanes, not one of them mentions travel? Or saving time? Or what a great business tool this machine can be? We get the idea that those are not really so important, and not the central reason that brings men and women into the sky. They talk, when we get to know them, of friendship and joy and of beauty and love and of living, of really living, firsthand, with the rain and the wind. Ask what they remember of their life so far and not one of them will skip the last twenty-three years. Not one.

"Well, right off the top of my head I remember chugging along there in formation with Shelby Hicks leading the way in his big Stearman biplane, heading for Council Bluffs, last month. And Shelby was flying and Smitty was in the front cockpit navigating—you know the way he does, real careful, with all his distances and headings just down to the exact degree—and all of a sudden the wind catches his map and pow! there it goes up and out of the cockpit like a big green ninety-mile-an-hour butterfly and poor Smitty grabs for it and he can't quite get it and the look on his face all horror and Shelby is sort of startled first and then he starts laughing. Even from flying alongside I can see Shelby laughing so there's tears running down inside his goggles and Smitty is disgusted and then in a minute he starts to laughin' and he points over to me and says, 'You're the leader!' "

A picture burned in memory because it was wild and fun and shared.

"I remember the time John Purcell and I had to land my plane in a pasture in South Kansas because the weather got bad all at once. All we had for supper was a Hershey bar. We slept under the wing all night, and found some wild berries that we were afraid to eat for breakfast at sunup. And ol' John saying my airplane made a lousy hotel because some rain got him wet. He'll never know how close I came to taking off and just leaving him out there in the middle of nowhere, for a while . . ."

Journeys across the middle of Nowhere.

"I remember the sky over Scottsbluff. The clouds must have gone up ten miles over our head. We felt like darn ants, I tell you . . ."

Adventures in a country of giants.

"What do I remember? I remember this morning! Bill Carran bet me a nickel he could take off in his Champ in less runway than I'd need in the T-Craft. And I lose, and I can't figure out why I lose because I always win with that guy, and just when I go to pay him I see he's snuck a sandbag into my airplane! So he had to pay me a nickel for cheating and another nickel for losing the takeoff when we did it again with the sandbag out . . ."

Games of skill, with sneaky tricks unplayed since childhood.

"What do I remember? What don't I remember! But I'm not about to go back and live it over. Too much still to do now." And an engine starts and the man is gone, dwindling down to nothing against the horizon.

You reach a point, I found, where you begin to know that a pilot does not fly airplanes in order to get somewhere, although he gets to many somewheres indeed.

He doesn't fly to save time, although he saves that whenever he steps from his automobile into his airplane.

He doesn't fly for the sake of his children's education, although the best geographers and historians in class are those who have seen the world and its history in their own eyes, from a private airplane.

He doesn't fly for economy, although a small used aircraft costs less to buy and run than a big new car.

He doesn't fly for profit or business gain, although he took the plane to fly Mr. Robert Ellison himself out for lunch and a round of golf and back again in time for the board meeting, and so the Ellison account was sold.

All of these things, so often given as reasons to fly, aren't reasons at all. They're nice, of course, but they are only by-products of the one real reason. That one reason is the finding of life itself, and the living of it in the present.

If the by-products were the only purpose for flight, most of today's airplanes would never have been built, for there are a multitude of annoyances cluttering the path of the lightplane pilot, and the annoyances are acceptable only when the rewards of flight are somewhat greater than a minute saved.

A lightplane is not quite so certain a piece of transportation as an automobile. In poor weather, it is not uncommon to be held for hours, days sometimes, on the ground. If an owner keeps his airplane tied down on the airport grass, he worries with every windstorm and scans every cloud for hail, much as if his airplane were his wife, waiting out in the open. If he keeps his plane in a large hangar, he worries about hangar fires, and thoughtless line-boys smashing other aircraft into his own.

Only when the plane is locked away in a private hangar is the owner's mind at rest, and private hangars, especially near cities, cost more to own than does the airplane itself.

Flying is one of the few popular sports in which the penalty for a bad mistake is death. At first, that seems a horrible and shocking thing, and the public is horrified and shocked when a pilot is killed committing an unforgivable error. But such are the terms that flying lays down for pilots: Love me and know me and you shall be blessed with great joy. Love me not, know me not, and you are asking for real trouble.

The facts are very simple. The man who flies is responsible for his own destiny. The accident that could not have been avoided through the action of the pilot is just about nonexistent. In the air, there is no equivalent of the child running suddenly from between parked cars. The safety of a pilot rests in his own hands.

Explaining to a thunderstorm for instance, "Honest, clouds and rain, I just want to go another twenty miles and

then I promise to land," is not much help. The only thing that keeps a man out of a storm is his own decision not to enter it, his own hands turning the airplane back to clear air, his own skill taking him back to a safe landing.

No one on the ground is able to do his flying for him, however much that one may wish to help. Flight remains the world of the individual, where he decides to accept responsibility for his action or he stays on the ground. Refuse to accept responsibility in flight and you do not have very long to live.

There is much of this talk of life and death, among pilots. "I'm not going to die of old age," said one, "I'm going to die in an airplane." As simple as that. Life, without flight, isn't worth living. Don't be startled at the number of pilots who believe that little credo; a year from now you could be one of them, yourself.

What determines whether you should fly, then, is not your business requirement for an airplane, not your desire for a challenging new sport. It is what you wish to gain from life. If you wish a world where your destiny rests completely in your own hands, chances are that you're a natural-born pilot.

Don't forget that "Why fly?" has nothing to do with aircraft. It has nothing to do with by-products, the "reasons" so often put forth in those pamphlets to potential buyers. If you find that you are a person who can love to fly, you will find a place to come whenever you tire of a world of TV dinners and people cut from cardboard. You will find people alive and adventures alive and you will learn to see a meaning behind it all.

The more I wander around airports across the country, the more I see that the reason most pilots fly is simply that thing they call life.

Give yourself this simple test, please, and answer these simple questions:

How many places can you now turn when you have had enough of empty chatter?

How many memorable, real events have happened in your life over the last ten years?

To how many people have you been a true and honest friend—and how many people are true and honest friends of yours?

If your answer to all these is "Plenty!" then you needn't bother with learning to fly.

But if your answer is "Not very many," then it just might be worth your while to stop by some little airport one day and walk around the place and find what it feels like to sit in the cockpit of a light airplane.

I still think of my salesman acquaintance of the airline flight between San Francisco and Denver. He had despaired of ever finding again the taste of life, at the very moment that he moved through the sky that offers it to him.

I should have said something to him. I should at least have told him of that special high place where a few hundred thousand people around the world have found answers to emptiness.

I've never heard the wind

Open cockpits, flying boots, and goggles are gone. Stylized cabins, air conditioners, and sun-shaded windshields are here. I had read and heard this thought for a long time, but all of a sudden it sank in with a finality that was disturbing. We have to admit to the increased comfort and all-weather abilities of modern lightplanes, but are these the only criteria for flying enjoyment?

Enjoyment is the sole reason many of us started to fly; we wanted to sample the stimulation of flight. Perhaps in the back of our minds, as we pushed the high-winged cabin into the sky, we thought, "This isn't like I hoped it would be, but if it's flying I guess it will have to do."

A closed cabin keeps out rain and lets one smoke a cigarette in unruffled ease. This is a real advantage for IFR conditions and chain smokers. But is it flying?

Flying is the wind, the turbulence, the smell of exhaust, and the roar of an engine; it's wet cloud on your cheek and sweat under your helmet.

I've never flown in an open-cockpit airplane. I've never heard the wind in the wires, or had only a safety belt between me and the ground. I've read, though, and know that's how it once was.

Are we doomed by progress to be a colorless group who take a roomful of instruments from point A to point B by air? Must we get our thrill of flying by telling how we had the needles centered all the way down the ILS final? Must the joy of being off the ground come by hitting those checkpoints plus or minus fifteen seconds every time? Perhaps not. Of course, the ILSs and the checkpoints have an important place, but don't the seat of the pants and the wind in the wires have their places too?

There are old-timers with frayed logbooks that stop at ten thousand hours. They can close their eyes and be back in the Jenny with the slipstream drumming on a fuselage fabric; the exhilaration of the wind rush through a hammerhead stall is there any time they call it up. They've experienced it.

It isn't there for me. I started to fly in a Luscombe 8E in 1955, no open cockpits or wires for us new pilots. It was loud and enclosed, but it was above the traffic on the highways. I thought I was flying.

Then I saw Paul Mantz's Nieuports. I touched the wood and the cloth and the wire that let my father look down on the men who fought in the mud of the earth. I never got that delicious excited feeling by touching a Cessna 140 or a Tri-Pacer or even an F-100.

The Air Force taught me how to fly modern airplanes in a modern efficient manner; no covering the airspeed indicator here. I've flown T-Birds and 86s and C-123s and F-100s. The wind hasn't once gotten to my hair. It has to get through the canopy ("CAUTION—Do not open above 50 knots IAS"), then through the helmet ("Gentlemen, a square inch of this fiberglass can take an eighty-pound shock force"). An oxygen mask and a lowered visor complete my separation from possible contact with the wind.

That's the way it has to be now. You can't fight MIGs with an SE-5. But the spirit of the SE-5 doesn't have to disappear, does it? When I land my F-100 (chop the power when the main gear touches, lower the nose, pull the drag chute, apply brakes till you can feel the antiskid cycle), why can't I go to a little grass strip and fly a Fokker D7 airframe with one hundred fifty modern horses in the nose? I'd pay a lot for the chance!

My F-100 will clip along at Mach One plus, but I don't feel the speed. At forty thousand feet, the drab landscape creeps under the droptank as if I were in a strictly enforced twenty-five mile speed zone. The Fokker will do an indicated one hundred ten miles per hour, but it will do it at five hundred feet and in open air, for the fun of it. The landscape wouldn't lose its color to altitude, and the trees and bushes would blur with speed. My airspeed indicator wouldn't be a dial with a red-line somewhere over Mach One, it would be the sound of the wind itself, telling me to drop the nose a little and get ready to hop on the rudders, for this plane doesn't land itself.

"Build a World War I airframe with a modern engine?" you ask. "You could get a four-place plane for the money!"

But I don't want a four-place plane! I want to fly!

I shot down the Red Baron, and so what

It was not a Mitty dream. It was no fantasy at all. That was a hard roaring black-iron engine bolted to the firewall ahead of my boots, those were real Maltese-crossed wings spanning out over my cockpit, that was the same ice-and-lightning sky I had known most of my life long, and over the side it was a long fall to the ground.

Now, down there in front of me, was a British SE-5 fighter plane, olive drab with blue-white-red roundels on the wings. He hadn't seen me. It all felt exactly the way I had known it would feel, from reading the yellow old war-books of flight. Exactly that way.

I stepped hard on the rudder bar, pulled the joystick across the cockpit and rolled down on him, tilting the world about me in great sweeping tilts of emerald earth and white-flour cloud, and blasting slants of blue wind across my goggles.

While he flew along unaware, the poor devil.

I didn't use the gunsight because I didn't need it. I lined the British airplane between the cooling jackets of the two Spandau machine guns on the cowl in front of me, and pressed the firing button on the stick.

Little lemon-orange flames licked from the gun muzzles with a faint pop-pop over the storm of my dive. Yet the only move the SE made was to grow bigger between my guns.

I did not shout, "Die, Englander pig-dog!" the way the Hun pilots used to shout in the comic books.

I thought, nervously, You'd better hurry up and burn or it'll be too late and we'll have to do this all over again.

In that instant a burst of night swallowed the SE. It leaped up into an agonized snap roll, clouting black from its engine, pouring white fire and oil smoke behind it, emptying junk into the sky.

I dove past him like a shot, tasting the acid taste of his fires, twisting in my seat to watch him fall. But fall he did not. Smoke gushing dark oceans from his plane, he wobbled half-turn through a spin, pointed straight down at me, and opened fire with his Lewis gun. The orange light of the gun barrel flickered at my head, twinkling in dead silence from the middle of all that catastrophe. All I could think was, Nicely done. And that this must have been just the way it was.

The Fokker snatched into a vertical climb in the same instant that I hit the switch labeled SOOT (*foof!* from beneath my engine) and the one next to it labeled SMOKE. The cockpit went dim in roiling yellow-black which I breathed in tiny gasps. Right rudder to push the airplane into a falling slide to the right, full back-stick to spin it. One turn . . . two . . . three . . . the world going round like a runaway Maytag. Then a choking recovery into a diving spiral, followed every foot by that river of wicked fog.

Presently the cockpit cleared and I recovered to level flight, a few hundred feet above the green farms of Ireland. Chris Cagle, flying the SE-5, turned a quarter mile away, rocking his wings in signal to join in formation and fly home.

As we crossed the trees side by side and touched our tail-skids to the wide grass of Western Aerodrome, I counted that this had been an eventful day. Since dawn I had shot down one German and two British airplanes, had myself been shot down four times—twice in an SE-5, once in a Pfalz, once in this Fokker. It was a lively introduction to the way that a movie pilot earns his keep, and there was a month more of it to come.

The film was Roger Corman's *Von Richthofen and Brown,* an epic featuring a fair amount of gore, some sex, a tampering with history, and twenty minutes of aerial footage that several living pilots nearly stopped living to produce. The gore and sex and history were make-believe, but the flying, as flying always is, was the real thing. Chris and I learned that first day in the air what every movie pilot since *Wings* has known: nobody has ever told the airplanes that this is all in fun. The aircraft still stall and spin, they'll have real mid-air collisions if you let them do it. No one else but pilots can understand this.

The camera tower was an excellent example. Our camera tower was a place built of telephone poles, a platform thirty feet above a knob of ground called Pigeon Hill. The cameraman and two assistants would climb to that platform every morning in sweet assurance that since this was only a film, they would live to climb down from the tower every afternoon. They had a trust in Chris and me and Jon Hutchinson and the dozen Irish Air Corps pilots that was beyond blind . . . the cameramen acted as if the aircraft diving down on them for head-on shots, guns blazing, were already pussycats safe on film.

It is ten a.m. We are a flight of two Fokker D-7s and two SE-5s. The engines and the wind are clattering about our heads and down there off our wingtops is the lonely lump of Pigeon Hill, with its tower on top and its cameramen on their platform.

"We want a tail chase this morning," they tell us on the radio. "An SE in front, a Fokker after him, another SE, and the other Fokker. You got that?"

"Roj."

"Come on close to the tower, please, then bank up on one wing and turn around us so we can see the tops of your airplanes. Close as you can to each other, please."

"Roj."

So here we go, from a thousand feet above the ground we fall into tight line-astern formation, the airplane ahead looming gigantic in our windscreen. Here's the dive at the tower, that tiny pyramid down there.

"Action! This is a take!"

The SE in the lead jinks violently back and forth, aiming for the tower and the ground. We follow him in the Fokker, firing short bursts of oxyacetylene from our fake guns, aware that another SE is close under our tail, firing, and that the other Fokker is under his. From moment to moment we catch the propwash of the plane ahead, which slams us up into a roaring bank that takes full opposite aileron and rudder to control. This is no problem, with room beneath us. But the room dwindles fast, and in a few seconds the camera tower is a pretty big thing, then a monster, and the cameraman is wearing a white shirt and a blue jacket and a red-and-blue scarf and the SE banks hard around the tower and we're in the WASH AND STICK RUDDER LOOKOUT WE'RE GONNA ROLL RIGHT INTO . . .

Gag. Ark. Foosh. We caught it in time the camera tower has flicked past and we're in one piece and man I thought we had had it then what a way to start off a day and oh boy this ain't fun this is WORK!

"All right. That was all right, chaps," comes the radio. "Let's try it again, and this time could you come a little closer in to the tower and don't get quite so far apart. Bunch it up a little bit more, please."

"Roj."

Dear God in heaven, he wants us CLOSER!

Down we come again line-astern, jinking, swerving, guns popping, close as we can force ourselves to dare, slamming in propwash that grabs us like a big hand and torques us, if we don't fight, all the way upside-down. The tower rises up at us like an Aztec pyramid of human sacrifice and then "SMOKE NOW, NUMBER ONE, SMOKE SMOKE!"

The SE we're chasing hits his smoke a hundred yards from the tower and it's like flying into the side of a thundercloud. The plane rolls wild left and we can't see a thing except a corner of blurred green that was the ground a second ago and we can't breathe and somewhere an instant away is the camera tower with those poor dumb trusting slobs cranking away with their little Mitchell, taking pictures. Stomp the right side of the rudderbar for dear life, snatch the stick back hard and we come blasting out of the smoke twenty feet left of the tower. We miss them by twenty feet. It's interesting to see how quickly a leather flying helmet can get soaked through with sweat.

"That was perfect. That was absolutely right. Now let's do that one more time . . ."

"ONE MORE TIME? REMEMBER THIS IS A HUMAN LIFE YOU'RE DEALING WITH!"

It was an Irish pilot who said that, and I remember thinking that his words were well said, my friend, well said.

I kept seeing, the more the tower called for closer and closer passes, that comedian who stands with a banana cream pie while the other one shouts, "Let me have that pie! Let me have it! LET ME HAVE IT!" The temptation

is to fly right straight down the center of that Mitchell, rip the thing to a billion pieces over the countryside, then pull up and say, "There! Is that close enough? Is that what you guys want?"

The only one who gave in to temptation was Chris Cagle. He came at the camera in anger, from below the tower, and climbed full throttle, splitting seconds, into the lens. Pulling up at the very last quarter instant, he got the grim pleasure of a millisecond view of the camera crew diving for the deck. That was the only time in the month that they thought the airplanes might be real, after all.

Most of the air-to-air photography in *Von Richthofen and Brown* was shot from a jet helicopter, an Alouette II. The helicopter cameraman wasn't visited with quite the same death-wish as the tower crew, but a helicopter is an unnerving thing to fly with. Just because the machine is pointed forward, of course, doesn't mean that it is moving forward—it could be stopped, or going straight up or down or backward. How does a pilot judge where to aim, to come a safe distance from an object of unknown velocity?

"OK. I am hovering," the pilot would tell us. "You can come in any time." But closing rate on a stopped helicopter is just the same as closing rate on a cloud, and that can be alarmingly fast, in the final seconds. One keeps thinking, too, that the poor souls inside the Alouette don't have parachutes.

Bit by harrowing bit, though, we made the film. We got used to the airplanes, for one thing. Most of the replicas did well to climb two hundred feet per minute after takeoff, and on some days were pressing their luck to clear the canvas hangars at the end of the field. In the immortal words of Jon Hutchinson, "I have to keep telling myself, 'Hutchinson, this is marvelous, this is lovely, you're flying a D-7!' Because if I don't, it feels like I'm flying a great bloody pig."

The four miniature SE-5s were not only at full power to stay with the other airplanes, they were at more than full power. On one flight I chased the Fokker Triplane with a camera mounted on the cowl of a mini SE, and just to stay in the same sky with the Fokker, eighty miles per hour, I was pulling 2650 rpm on an engine red-lined at 2500. Out of that fifty-minute flight, forty-five minutes were spent on the other side of full throttle. The film, like a war, was a mission that had to be accomplished. If an engine blew up that was just too bad . . . we'd have to land somehow and take up another airplane.

Odd, but one gets used to this kind of flying. In time, even on the tower at Pigeon Hill, caught in propwash and rolling out of control thirty feet in the air, one thinks, I'll save it. She'll recover at the last second. She always has . . . all the while pouring the power of Charles Atlas into the controls, fighting to pull out.

One day I saw an Irish pilot all alone, wearing a sprig of heather in the lapel of his German flying jacket.

"Flying kind of low, aren't you?" I said, by way of a joke.

His face was gray; he didn't smile at all. "I thought I had had it. I am lucky to be alive."

It was such a somber voice that I was caught in morbid curiosity. The leaves in his lapel came from the downslope at Pigeon Hill, and he had harvested it with the undercarriage of a Fokker.

"The last thing I remember was the propwash and all I saw was the ground. I closed my eyes and pulled hard as I could on the stick. And here I am."

The tower crew confirmed it that evening. The Fokker had rolled and dived as it passed the tower, bounced off the side of the hill and back into the air. The camera was pointed the other way.

One of the airplanes at Weston was a two-seater, a Caudron *277 Luciole,* which was translated for us as *Glowworm.*

It was a square sluggish biplane with a Lewis gun mounted in the rear cockpit in such a way that there was not quite enough room for the gunner to wear a parachute. Hutchinson, just down with the machine as I was about to take it up, described it for me in his pure British tones: "It's a fine luciole, actually, but it will never be an airplane."

Thinking that over, I fastened myself into the front seat, started the engine, and took off for a mission in which I was to be shot down by a pair of Pfalzes. It was not an enjoyable scene at all. It was much too real.

The poor Caudron could barely stumble out of its own way, much like the great majority of real two-seaters of the First War. It could neither turn nor climb nor dive, and the pilot sits directly between the wings so that he cannot see up and he cannot see down. The gunner blocks the view behind and the pilot gets what's left over: a slice of sky ahead, and, sieved through the struts and wires, to the side.

I thought I had understood that two-seater pilots lived a hard life in 1917, but I hadn't understood that at all. They couldn't fight, they couldn't run away, they could hardly tell that they were being attacked until their little fabric coffin burst into flames and then they didn't have parachutes to bail out with. Perhaps I was a two-seater pilot in another life, for in spite of myself, in spite of saying, "This is a movie, Richard, this is only a movie that we are taking pictures for," I was frightened when the Pfalzes came in. Their guns sparkled at me, the director shouted, "SMOKE, LUCY, SMOKE, SMOKE!" I hit both smoke switches, slumped in the seat, and wallowed the *Luciole* into a low-speed spiral dive.

That was the end of the scene for me, simple as that, but I dragged back to Weston like an exhausted snail.

Turning downwind to land, I suddenly saw a flight of Fokkers turning toward me, and went cold in shock. It took seconds to remember that this was not 1917 and that I was not going to be incinerated in my own traffic pattern. I laughed, then, nervously, and got the airplane on the ground as fast as I could. I had no wish to fly the two-seater again and I never did.

Nobody was killed in the time I flew with *Von Richthofen and Brown;* nobody was even injured. Two airplanes were damaged: an SE with an axle failure while taxiing, a Pfalz in a groundloop. Both were flying again within a week.

The cameras rolled through thousands of feet of color film, hours of film. Most of it looked pretty tame, but for every time that a pilot was truly frightened, certain that he was going to be a mid-air collision, positive that this time the plane was not going to recover at low altitude, there was another exciting scene caught in celluloid.

We gathered in tight little knots to watch the previous day's action on the six-inch screen of the Movieola. No sound save the whir of the projector; quiet as a small-town library. Occasional comments: "Move it in!" "Liam, was that you in the Pfalz?" "That's not too bad, there . . ."

As the filming went into the final week, painters converged on the drab German airplanes and brushed them into the flying rainbows of the Richthofen Circus. We flew the same airplanes as before, but now it was a point of fun to fly the all-red Fokker that would appear on the screen as Von Richthofen himself, or the black Pfalz that would be Hermann Goering's.

I drew the red Fokker once for the ignoble scene of having one of my wingmen shot down by the Englander. Then once again as the Red Baron to come roaring to the rescue of Werner Voss, shooting an SE off his tail.

The next day I was Roy Brown, chasing Von Richthofen (a red Fokker Triplane, now) and shooting him down for the final scene of the film.

I tried saying it when I climbed out of the cockpit after that flight, carrying my parachute through the quiet evening to our trailer. "I shot down the Red Baron."

I thought about that. How many pilots can make that statement? "Hey, Chris," I said. He was stretched out in his half of the trailer. "I shot down the Red Baron!"

His reply was incisive. "Hm," he said. He didn't even open his eyes.

Which was to say, So what? So it's just a movie we're flying for, and a B movie besides and if it wasn't for the flying scenes, I wouldn't cross the street to see the picture, at home.

That's when it occurred to me that it's the same in a real war as it was in ours of make-believe. Pilots don't attend wars or films because they like the blood or the sex or the B-level plots of the things. More important than film is the flying; more important than war is the flying.

It's probably a shame to say: neither films nor wars will ever lack for men to fly their airplanes. I am myself one of a great many who volunteered for both. But surely someday, a thousand years from now, we can build a world where the only place to log combat time is in the lens of some director shouting, "SMOKE NOW, SMOKE!"

All we need is the will to do it, some replica MIGs, some antique Phantoms with dummy guns, sawdust missiles . . . If we wanted to, a thousand years from now, we could really make some great films.

Prayers

"You'd better be careful what you pray for," somebody once said, "because you're going to get it."

I thought of that, twisting a Fokker D-7 hard through my little part of the Great Mass Dogfight scene in *Von Richthofen and Brown.* The scene had looked neat and safe when we chalked it out on the briefing-room blackboard, but now, in the air, it was scary—fourteen replica fighters crushed into one small cube of sky, each one chasing the other, a few losing position and diving blindly through the rest, rainbow paints flashing colored sunlight, the loud quick blast of a Pfalz engine as the plane flashed beneath without seeing, smoke trails and the thick smell of fireworks in the wind.

Everyone survived that morning, but I was still shaking a bit when I thought about being careful what we pray for. Because the very first magazine article I wrote, twelve years ago, was one in which I prayed that those of us who learned to fly in closed-cockpit airplanes might have a place to rent an open-cockpit one, for the fun of it, ". . . and fly a Fokker D-7 airframe with one hundred fifty modern horses in the nose," I had written. And here I was this moment in helmet and goggles and scarf, pilot of a yellow-blue-white-green airplane, *Fok. DVII* lettered authentically on the fuselage. I came home from the film with forty hours in Fokkers and Pfalzes and SE-5s, my prayers answered so completely that I had all that kind of flying I cared to do for quite some time.

A few years after I had prayed for the Fokker, I had

gone for a ride in Chris Cagle's J-3 Cub, at the Merced Fly-in. Cagle had a thousand hours in that Cub alone, I guess, and as we flew across the afternoon he showed me how to fly at zero miles per hour and how to loop and roll the thing. I remember looking out the open door at the puffed yeast-doughnut tire, and past it to the ground way down below, thinking what a great little airplane, and some day, by God, I'll own me a Cub! Today I own it, and it has big puffy yeast-doughnut tires and the doors open in flight and I look down and remember, Sure enough, it happened again: I got what I prayed for.

Time after time I've watched it happen, in my life and the lives of people I know. I've tried to find somebody who *didn't* get what he prayed for, but to date I haven't found him. I believe it: whatever we wrap away in thought is opened for us, one day, in experience.

There was a girl I met in New York, who lived in a tight-packed Brooklyn tenement, acred about by old concrete and cracking brick, by frustration and fear and quick wild violence in the street. I wondered aloud why she didn't get out, move to Ohio or Wyoming country, where she could breathe free and touch the grass once in her life.

"I couldn't do that," she said, "I don't know what it's like out there." And then she said a very honest and knowing thing. "I guess I'm more afraid of what I don't know than I hate what I have right now . . ."

Better to have riots in the streets, better squalor and subways and sardine crowds, she prayed, than the unknown. As she prayed, she received; she meets nothing now that she hasn't met before.

All at once I saw the obvious. The world is as it is because that is the way we wish it to be. Only as our wish changes does the world change. Whatever we pray for, we get.

Look about, sure enough. Every day the footsteps of answered prayer are ours to walk, we have only to lean forward and walk them, one by one. The steps to my Fokker were many. I helped a man with his magazine, years ago, and so came to know him. His prayers were in old airplanes and business deals and motion pictures, and he took his chance to buy, in a business deal with a film studio, the fleet of World War I fighters. When he mentioned this, I said I'd be ready if he ever needed a pilot to fly one; that is,

I took one step that offered itself to be taken. A year later he needed two American pilots to join the group, in Ireland, flying the Fokkers. When he called, I was ready to finish the path I had begun with the first article, that first prayer about the D-7.

From time to time, when I was barnstorming the Midwest a few summers ago, a passenger or two would say, "What a great life you have, free to go wherever you want, whenever . . . Sure wish I could do it." Wistful, like that.

"Come along, then," I'd say. "You can sell tickets, keep the crowds behind the wing, strap the passengers into the front seat. We might make enough money to live on, we might go broke, but you're invited." I could say this, first because I could always use a ticket seller, and second because I knew what the answer would be.

Silence first, then, "Thanks, but you see, I've got my job. If it wasn't for my job, I'd go . . ." Which was only to say that each wistful one wasn't wistful at all, each had prayed harder for his job than for the life of a barnstormer, as the New York girl had prayed more for her tenement than for the grass of Wyoming or for any other unknown.

I consider this from time to time, flying. We always get what we pray for, like it or not, no excuses accepted. Every day our prayers turn more into fact; whom we most want to be, we are. It all sounds like justice to me; I can't say as I mind the way this world is built, at all.

Return of a lost pilot

We had been flying north, low-level formation in a pair of F-100 day fighters out over the Nevada desert. I was leading, that time, and Bo Beaven's airplane was twenty feet away at my right wingtip. It was a clean morning, I remember, and we were cruising three hundred feet above the ground. I was having some trouble with the radiocompass, leaning down in the cockpit, resetting a circuit breaker, clicking the control from ANT to LOOP to COMP, to see if the needle would show any life. Then about the time I thought that the problem was in the antenna itself, and that maybe I shouldn't plan on having any help from the radio at all, there came Beaven's voice filtered in my earphones. It was neither a command nor a warning . . . it was a simple calm question: "Do you plan on flying into this mountain?"

I jerked my head up, startled, and there angled in front of us was a rugged little mountain, all brown rock and sand and tumbleweed, tilting, flying toward us at something over three hundred nautical miles per hour. Beaven said nothing more. He didn't loosen his formation or move to break away. He spoke in the way that he flew his airplane . . . if you choose to fly straight ahead, there will be not one hole in the rock, but two.

I eased the control stick back, wondering where the hill had come from, and it flicked a hundred feet beneath us and was gone, silent as a deadly dark star.

I never forgot that day, or the way Beaven's airplane faced the mountain wing to wing with mine, not clearing the peak until we cleared it together. It was our last flight in formation. A month later our time had run out in the peacetime Air Force and we were civilians again, promising, sure, we'd meet again, because people who fly always meet again.

Back in my home town, I was sad to be gone from high-performance flying only until I found that the same tests waited in lightplane sport flying. I discovered formation aerobatics, air racing and off-airport landings, all in little planes that can take off and land five times in the distance it takes an F-100 to get off the ground once. I thought, as I flew, that Bo would be making the same discovery, that he was flying just as I was.

But he wasn't. He was no sooner out of the Air Force than he was lost, no sooner established in business than he was dead, the agonizing death of the pilot who turns his back on flight. He suffocated slowly, the blue-suited businessman had taken over, had mortared him into an airless corner behind a wall of purchase orders and sales charts, golf bags and cocktail glasses.

Once, on a flight through Ohio, I saw him long enough to be sure that the man who controlled his body was not the same man who had flown my wing that day toward the mountain. He was polite enough to recognize my name, to wish me good day, but he heard without interest any talk of airplanes, wondered why I looked at him strangely. He insisted that he was indeed Bo Beaven and quite happy as an executive for a company that made wringer washers and plastic products. "There's a great demand for wringer washers," he said, "a lot more than you might think."

Way far down in his eyes I fancied I saw a faint little signal of despair from my friend trapped within, fancied I heard the smallest cry for help. But it was gone in a second, quickly masked by the businessman at the desk, behind the nameplate *Frank N. Beaven*. Frank!

It used to be, when we were flying, anybody who called Bo by the name "Frank" advertised he was no friend at all. Now the clumsy business executive had made the same mistake; he had nothing in common with the man he had sealed up to die.

"Of course I'm happy," he said. "Oh, sure, it was fun to fly around in the '100, but that couldn't go on forever, could it?"

So I flew away and Frank N. Beaven went back to work

at his desk, and we didn't hear from each other again. Maybe Bo had saved my life with his cool question in the desert, but when he needed me to save his, I didn't know what to say.

It was ten years from the day we had left the Air Force, then, that I got a note from Jane Beaven. "Thought you'd be pleased to know that Bo made his move and is at last returning to number one love, the flying business. With American Aviation in Cleveland—is like a new man . . ."

My friend Bo, I thought, forgive me. Sealed away for ten years and now you come crashing through the wall. You're a tough one to kill, aren't you?

Two months later I landed at Cuyahoga County Airport, Cleveland, and taxied to the American Aviation factory, with its pond of bright-painted Yankees awaiting delivery. And out across the ramp came Bo Beaven to meet me. He wore white shirt and tie, to be sure, but it was not the businessman Frank, it was my friend. There were just bits of the Frank-mask left about him, bits that Bo had allowed to remain because they served a purpose in his job. But the man who had been walled away from the sky was now alive and well and in full charge of the body.

"You wouldn't have any of these planes to deliver east, would you?" I said. "Maybe you and I could ferry one out."

"Who's to say? We just might have one to go." He said it with a perfectly straight face.

His office now is the office of the Director of Purchasing, a mildly cluttered place with a window overlooking the factory floor. There on a filing cabinet stands a scratched and battered company model of an F-100, pitot boom missing, decal shredded, but proud and there, banking into the indoor sky. On the wall is a photograph of a pair of Yankees in formation over the Nevada desert. "That look familiar?" he asked shortly. I didn't know whether he meant the desert or the formation. They were both familiar, to me and to Bo; the businessman Frank had never seen either one.

He showed me around the Yankee plant, at ease in this place where the seamless sport plane comes to life out of metal as he had come to life out of grounded flesh. He talked about the way the Yankee is bonded together instead of riveted, about the strength of the honeycomb cabin section, about problems in sheet planning and the shape of a control wheel. Technical business talk, for sure, but the business now was airplanes.

"All right, fella. What was it like, what has it really been like for you, the last ten years?" I said, relaxing in the car while he watched the road home carefully, not looking at me.

"I used to think about it," he said slowly, "the first year out of flying, wandering to work in the morning when there was a bit of cloudiness. I'd think of the sun, up on top. It was awfully hard." He took the turns fast, keeping his eyes on the road. "The first year was bad. But by the end of the second year, I almost never thought about it; but occasionally I would maybe in the corner of my ear hear an airplane above an overcast or something, and give some thought. Or maybe for business reasons I'd take a commercial flight to Chicago and have occasion to go on top, and then I'd remember all these things. 'Yes, I used to do this frequently, that was fun, that was enjoyable, that made you feel clean and all that sort of stuff.' But then I'd land, get to the business of the day, and maybe sleep on the way back, and I wouldn't have that thought, I wouldn't think of it tomorrow, or the next day."

Tree-shadows flickered over the car. "I was unhappy, with that company. It had no relation to a product that I knew about or was interested in. I didn't care if they ever sold another wringer washer or another ton of reclaimed rubber or another carload of diaper pails. I didn't care at all."

We stopped at his house, a white-painted lawn-surrounded picket-fence place in the shade of Maple Street, Chagrin Falls, Ohio. It was a moment before he left the car.

"Don't get me wrong, now. I don't think that at any time, other than just flying alone, tooling around, did I ever give any thought to things like breaking through the overcast. When I saw the sun, it was what I expected to see. It was very nice, pleasant to see all the clean tops-of-clouds where underneath there were all the dirty bottoms-of-clouds. But I don't think I had any lofty godly-type thoughts when I was flying, that sort of thing.

"It might have been very casual, I might have broken out and said mentally, 'Well, God, here I am up here looking at it the way you're looking at it.' And God would say, 'Roj,' and that'd be all there was to that. Or he'd click his mike button to acknowledge that I had spoken.

"I was always awe-stricken at how much there was of the top of clouds. And the fact that I was up there, tooling around with the bigness of it all, skirting a big thunderhead or something like that, when people on the ground were merely deciding whether they should take their umbrella. I'd think of these things, wandering to work . . ."

We walked to the house, and I tried to remember, No, he had never talked that way, he had never said that kind of thing out loud, as long as I had known him.

"And now," he said after supper, "well, very few people know of American Aviation. They either don't know it, or they screw it up and say, 'Oh, that's the operation that's going broke, or went broke.' That's good, because then I can give them my speech: 'No, this isn't going broke, this is American Aviation. We've got people who are pros . . .' and all this sort of thing. And they *are* pros. This is one of the other things I wanted to do when I quit the wringer-washer job—I didn't want to work with a bunch of . . . well I wanted to work with a more professional organization."

We checked the Yankee for its ferry flight to Philadelphia, and I remembered what Jane Beaven had said the day before. "I don't know him and I never will. But Bo was a changed man, when he went completely away from flying. It got to him, he was understimulated, he was bored. He doesn't talk a great deal about what he feels, he doesn't go on and on about anything. But when he quit at last, he had two choices of excellent jobs. One was with a big metals company and he would be there forever, and the other was with American Aviation which could, as far as we really knew, fold the next day. But after one interview I knew where we were going." She had laughed out loud. "He kept saying, of course, 'The metals company would be marvelous, and much more secure,' and all that, and to me it was the biggest line of hogwash . . . I knew where we were going."

The Yankee rolled out onto the runway, one of Beaven's first flights after his years on the ground. "You've got it, Bo," I said. "Your airplane."

He pressed full throttle, tracked the centerline, and we found that the Yankee, over grass, on a hot day, is not a short-field airplane. We left the ground a good way down the runway, angling long and shallow up into the air.

The ten years absence showed, even in a man who at one time had been a better pilot than I could hope to be. He wasn't thinking ahead of the airplane, he was rough on the controls, and the sensitive little Yankee pitched and rolled under his hands.

But oddly, he was perfectly confident. He was rough and he knew it, he was behind the plane and he knew it, but he also knew that all this was normal as he got used to flying again, and that he'd catch up before many minutes had gone by.

He flew the Yankee the way he last remembered to fly; he flew it like a North American F-100D. Our turn on course wasn't a gentle sweeping general-aviation turn, it was WHAM! the wing slammed into a steep bank, dug into the air, *turned,* then flew back to wings-level in a furious hard whiplash.

I had to laugh. For the first time I could see what another human being saw, I could look inside his mind. And I saw not a little civilian Yankee slicing along at one hundred twenty-five miles per hour with a hundred horsepower spinning a fixed-pitch propeller up front, but a D-model F-100 single-seat day fighter streaking ahead of fifteen thousand pounds of thrust blasting diamond lights out the afterburner and the ground blurring by beneath us and that button-studded control stick under his hand, that magic grip that one need only touch to spin the world, or turn it upside-down or make the sky go black.

The Yankee didn't mind the game, for its flight controls very nearly match the '100's. The wheel is light and positive as a racing Ferrari's, so that one is tempted to fly hard fast eight-point rolls, just for fun.

Bo discovered the sky he had once known so well. "Will we ever own an airplane?" Jane had said. "I hope so. Because he'd fly. I can't explain to you why, because the inner workings of his mind are always his own, but I think he feels better, I think he feels more like living . . . this sounds very corny, but I think his life means more to him when he can fly." It didn't sound corny to me at all.

Bo squinted into the horizon. "Looks like the clouds are going broken, here. What do you say, over or under?"

"You're flying the airplane."

"Under."

He chose that for the fun of coming down. Carb heat and throttle, the Yankee snapped its wings up like a daylight bat, and we flashed down toward the trees. Bo was thinking ahead of the airplane now, and happy, though of course he didn't smile. The wings lashed level and we shot above the Pennsylvania Turnpike, heading eastward.

"He's a little bit afraid to let loose and commit himself completely," Jane had guessed about him. "He's a little bit leery to become again so totally involved as he was with airplanes before. He won't let himself go. But there's one thing about Bo. He doesn't have to use a lot of words. He can communicate with flying."

Right you are, Jane. It was there all around as he flew, ten years of standing on the ground wanting to shout, now that the time had come to fly again, and his pain that our mission was just to deliver this airplane straight and level to Philadelphia, instead of taking it there in loops and slow rolls. He didn't have to say a word.

"What do you remember about instrument flying?" I asked.

"Nothing."

"OK, then, you're on the gages. I'll be approach control. 'Four niner Lima in radar contact, climb to and maintain three thousand five hundred feet, turn right heading one two zero degrees, report crossing the one six zero degree radial Pottstown VOR.' " I had meant to bury him in instructions, but it didn't work. All I had given him was a target to shoot for, and he aimed and shot, offering no excuses. The Yankee climbed and turned smoothly now under his hand, it leveled, and he remembered out loud.

"A radial is always outbound from the station, isn't it?"

"Yep."

He called, crossing the radial.

So I was around to watch my friend learn again, to watch the sky blast dust and cobwebs from a man who had been a magnificent pilot and who just might be one again.

"I'm joining the Yankee flying club," he had told me. And another time, "It wouldn't be too expensive, would it, to get a Cub or a Champ, just to fly around in? And as an investment, of course; the way prices are going up it would probably be a good investment."

We dropped into the pattern at the 3M airport, and there it was again, I was watching through his eyes, and there was the smooth silver nose in front of us, and the arrow of the pitot boom, and we were smoking down final approach at one hundred sixty-five knots plus two knots for every thousand pounds of fuel over a thousand and speed brakes out and gear down and flaps down and trim . . .

The J-57 of the F-100 thundered soft in our ears, eighty-five percent rpm on final, hold the sink rate, antiskid on, stand by to deploy the drag chute. We touched, the two of us, in a 1959/1969 F-100/Yankee in Nevada/Pennsylvania, USA.

Then he pulled the nose up, after touchdown, way too high up, so that we nearly scraped the tailskid. "Bo, what are you doing?" I had forgotten. We pulled the nose up high, in those days, for aerodynamic braking, to slow the plane and save a drag chute. Of course he had forgotten, too, why anybody would want to pull the nose of an airplane up after touchdown.

"What a lousy landing," he said.

"Yeah, that was pretty grim. I don't know whether there's hope for you or not, Bo."

But I did have hope. Because my friend, who had saved my life, and then been dead himself for so long, was flying. He was alive again.

Words

We were fifty miles northwest from Cheyenne, level at twelve thousand five hundred feet. The Swift's engine hushed along up front as it had for three hours since take-off and as I hoped it would for another thirty hours of cross-country flying. The instruments were relaxed and content on the panel, touching pressures and temperatures and metals and airs and telling me that all was well. Visibility was unlimited. I had not filed a flight plan.

I was just up there flying along, thinking about semantics, without the faintest premonition of what was to happen in four and a half minutes. Looking around at the mountains and the high desert and the altitude and the oil pressure and the ammeter and the first few scattered clouds of the day, and thinking about some of the words of aviation, and what they mean to the rest of the world.

About *flight plan,* for instance. To thinking people a flight plan, obviously, is a plan for a flight. A flight plan is a certain order, a discipline, a responsibility to move with purpose through the sky. Flying without a flight plan, to any rational person, is flying without order, discipline, responsibility, or purpose.

Oil temperature seventy-five degrees Centigrade . . . it's a good feeling, to have that forward-mounted oil cooler, on a Swift.

But to the Federal Aviation Administration, I thought, a flight plan isn't a plan for flying at all. It is an FAA Form 7233-1. A flight plan is a five-by-eight-inch piece of paper which is filed to alert search and rescue when an airplane is overdue at its destination. To those who know, a flight plan is a piece of paper. Those who do not know believe that a flight plan is a plan for a flight.

I considered that, cruising west of Cheyenne. I remembered the news reports that I had read: "Today an airline jet transport taxied over a light Cessna training plane, parked and tied down at the airport. The Cessna, which was squashed flat, and had not filed a flight plan . . ."

Had not filed a flight plan, in news parlance, means, *Guilty. Cause of accident. Deserved everything he got.*

Why has the FAA never defined *flight plan* for news reporters? Is it because the Administration wants them to believe that anyone who has not requested search and rescue service on Form 7233-1 is guilty, and the cause of any accident? Strange how convenient it is, at the moment of any incident, to mention to reporters that the light aircraft was not on a flight plan. Or better, when they ask, "Did the little plane have a flight plan?" to reply reluctantly, with pain, "Well, gentlemen, no. Much as we hate to say it, the light aircraft had not filed a flight plan."

It was not just two minutes till the event-of-which-there-were-no-premonitions was to happen. Engine instruments steady. Heading 289 degrees. Altitude 12,460 feet. But I kept thinking about words. There are so many of them, so many labels and terms so carefully chosen by officials that suspicious pilots might almost think they were cunningly set snares for the private citizen who has learned to fly.

Control tower. Air traffic controller. Where did those names come from? They control nothing at all. The people in that tower talk to airplane pilots, advise them of conditions. The pilots do every bit of controlling that's done. A semantic detail, that, and of no importance? How many times have you heard nonfliers say, "Your airport has no control tower? Isn't that *dangerous?*" Imagine how they feel when they find that the official terminology for a no-tower field is *uncontrolled airport!* Try explaining that to a news reporter! The words alone show an accident waiting to happen, airplanes trembling to fall out of the sky onto schools and orphanages. Here is a description of millions and millions of takeoffs, the kind of takeoff made every day, every minute: *The light aircraft took off from an uncontrolled airport, without radio control, without a flight plan.* Wow.

Airway sounds like *highway,* a smooth place on the ground where automobiles move swiftly and efficiently. In fact, an airway is a channel forcing airplanes to fly as closely to each other as possible in what would otherwise be a limitless sky.

Quadrantal altitude. A very technical authorized term to describe a system that at its very best assures that every mid-air collision will occur at an angle of less than 179 degrees.

Look around for other airplanes. It's just too simple. In any society that refuses to trust a human being, in any civilization that requires guaranteed safety from infallible tin boxes instead of individual care, *look around* is embarrassingly undignified. Why, it's unsophisticated, that's what it is.

My time was up. I flew at exactly 12,470 feet, thirty feet below prescribed quadrantal altitude for westbound flights. I was on Victor 138, the airway from Cheyenne to Medicine Bow, Wyoming.

The other airplane was also on Victor 138, also at 12,470 feet, but it flew in a direction that would take it head-on through the spinner of the Swift, through the cockpit and aft fuselage, thence through to the rudderpost and the clear air beyond. The other aircraft was thirty feet below what was exactly the wrong altitude. I had the right of way, but he had the C-124, which was at one time the largest four-engine transport in the world.

The Swift and I decided not to argue about rights, and turned gently out of the way. The '124, we saw, is actually a very large airplane indeed.

I was astonished. Why, that man is a professional pilot, an Air Force pilot! And he's at MY altitude. He's at the wrong altitude! He's eastbound at the westbound altitude. How can a professional pilot, how can he possibly be so wrong, in such a gigantic airplane?

It wasn't a near miss. The '124 is a sufficiently monstrous chunk of iron to be seen long before near-miss time. But still, there it was, dead on my altitude, a hundred tons of aluminum-steel, going the wrong direction.

Had I been involved in an overlong session with my map, and had the giant in fact vaporized the Swift, no doubt exists as to the report that would appear in the news. After explaining that the Swift had been smashed to powder against a minor wing fairing of the transport, and perhaps showing the small dent that we would have made there, the news would have concluded like this: "FAA spokesmen expressed regret over the incident, but did admit under questioning that the light airplane had not filed a flight plan."

Across the country on an oil pressure gage

Do you ever get the feeling that everybody else knows something you don't know? That everybody else in the world is taking for granted something you haven't even heard about, as if you missed the Big Briefing In The Sky or something?

One of the primal points covered in the Big Briefing apparently was that People Don't Fly Old Airplanes From Coast To Coast. People In Their Right Minds, that is. Then along comes old Bach, who missed the Briefing.

The airplane that I wanted was a 1929 Detroit-Parks P-2A Speedster open-cockpit biplane, and it was in North Carolina. I wanted to trade my Fairchild 24 for it, and I was in California. Now doesn't it seem the most logical thing in the world to fly the Fairchild to North Carolina, pick up the biplane, and fly it to California? If that sounds logical, you missed the Briefing too. There's always us two percent who never get the word.

Therefore, not knowing any better, I flew my gentle, smooth-purring, instrument-humming cabin monoplane to Lumberton, North Carolina, and traded it for a ratchety, roaring, snarling, windy biplane whose only reliable instrument was an oil pressure gage, that had never heard of an electrical system, let alone a radio, and was extremely suspicious of any pilot who did not learn to fly in a JN-4 or an American Eagle.

Also discussed at the Briefing, I'm sure, was You've Got To Be A Mighty Good Aviator To Land Old Biplanes In A Crosswind On A Hard-Surface Runway. Which explains why suddenly there I was at Crescent Beach, South Carolina, listening to a strange scrunching, tearing sound as my groundloop collapsed the right main landing gear, demolished the right wheel, and turned the right lower wing into a frayed and haggard pretzel. Later I listened for a while to the distant roar of the Atlantic Ocean, and later still, after dark, to the tin pelting of sad rain on the hangar into which my wreckage-pile had been towed. And there were only twenty-six hundred miles to go. I longed for hemlock to drink, or a bridge from which to throw myself into the sea. But we who missed the Briefing are so helpless and deserving of pity that we somehow manage to crawl through life despite our shortcomings. Pity, in this case, came from the former owner of the Parks, by name Evander M. Britt, custodian of an unquenchable fount of southern hospitality. "Now don't worry, Dick," he said when I called. "I'll be down right away with a set of new landing gear. There's an extra wing here if you want it, too. Don't you worry. I'll be right down."

And with him, driving through the rain, Colonel George Carr, barnstormer, fighter pilot, squadron commander, antique airplane restorer. "Is that all that's wrong!" Carr said when he saw the wreckage. "From what 'Vander told me, I thought you had hurt something! Help me with this jack, and we'll have you in the air tomorrow."

The comfortable web of the Antique Airplane Association closed about its member in distress, and from Gordon Sherman, president of the Carolinas-Virginia Chapter, as from the Celestial City itself, came a rare old wheel from

his Eaglerock for my right main gear. In a few days the Parks and I were as good as the day we rolled from the factory, and having learned some lessons about the mixture of crosswinds and hard-surface runways, we gave humble thanks to our benefactors, accepted a survival-ration package from Colonel Carr, and began to nibble on the twenty-six hundred miles.

We nibbled, too, at thirty-five years, and I discovered that the pioneer barnstorming pilots who flew the Parks and her sister ships were the oiliest and the coldest men of their time. I discovered that firsthand. After each day's flying, in field or on airport, out comes a grease gun to force the sticky stuff into each rocker-box housing. Five cylinders, ten rocker-boxes. After each flight, out comes a rag to wipe the rocker-box grease from where it has been thrown onto everything behind the engine: goggles, windscreens, fuselage, landing gear, stabilizers. Wipe it off quickly, before it hardens. The Wright J-6-5 Whirlwind is an oily little personality itself, and opening the cowl to strain the fuel each morning marks the barnstormer with a tenacious film, the print of his calling.

I had known, of course, from reading my air temperature gages in airplanes past, that the higher one flies, the colder grows the air. But I discovered that looking at COLD on a gage and having it smash and twist through the cockpit, slicing through leather coats and woolen shirts, are two very separate and distinct experiences. Only by ducking well forward under the windscreen could I avoid the thun-

dering icy knives of a hundred-mile wind, and ducking well forward for three hours at a time can be less than comfortable.

I discovered a great and basic fact early in my acquaintanceship with the Parks, as I flew westward with the first days of spring, 1964. One enjoys the land over which one flies in direct proportion to the speed with which he flies over it. Caught in headwinds over Alabama meadows, I saw for the first time that each tree in spring is a bright green fountain, gushing brilliant leaves into the sun. Some of the pastures are like the rolled greens of the most exclusive country clubs, and it was all I could do to keep from landing in them for the sheer fun of rolling along the bright untouched grass. The Parks wasn't at all convinced that I was worthy to be her pilot, but from time to time she would show me these views of her world, views of What It Was Like Then. Farm after weathered farm sifted by, each reigning at the end of its own dirt road, guarding its fields and forests just as it did when the Parks was new and seeing it all for the first time herself. More than one farm drive harbored even the 1930 automobiles and trucks, pastures grazed 1930 cows, and I was for a moment the cold and oily Buzz Bach, helmeted and goggled barnstormer of the untrodden skies. It was so good an illusion that it was true.

But as I looked away for a moment to write a note on the corner of my road map, the Parks showed unhidden jealousy. Roaring along straight and level, I glanced aside and wrote "trees are green fountains" on the map. By the time my pencil point was finishing ". . . ns," the engine roar was much louder than it had been and the wind was screaming in the wires. I jerked my head up to see a great tilted earth rushing to crush me, and to hear a little soft voice say, "When you fly me, you must fly me, and not take notes or think of other things . . ." And sure enough, the Parks was impossible to trim to fly hands-off, and try as I might, she would invariably roll into a wild unusual attitude whenever I thoughtlessly diverted my attention from her needs.

Hours blended and ran together into long days of flying as the face of the southern United States rolled along beneath me. Three hours of flying were enough to cover the front cockpit windscreen with oil and rocker-box grease, but the Whirlwind's five cylinders thundered right along and didn't miss a single beat.

The Parks taught me something about people, when she judged me ready to learn. Get away from the cities, she said, and people have time to be outgoing and friendly and terribly kind. Take a little place like Rayville, Louisiana. Land on the little strip there as the sun is going down. Taxi to a short row of hangars, a fuel pump. Deserted, all. Shut down the engine, by a sign that says Adams Flying Service, with a Grumman Ag-Cat and a Piper PA-18 sprayer tied outside. Get out of the cockpit and stretch and start wiping rocker-box grease. And suddenly there's a pickup truck and a voice. "Hi, there."

The truck has Adams Flying Service painted on its door, and the driver is smiling and wearing an old felt hat with the brim turned up in front.

"Thought you were a Stearman at first when you went over my farm, but you were too little to be a Stearman and that didn't sound like any 220 engine. What kind of airplane is that, anyway?"

"Detroit-Parks. Just like a Kreider-Reisner 34, if you know that airplane."

The talk started about airplanes and the man was Lyle Adams, owner of his own crop-dusting company, once a bronc-rider, bulldogger, charter pilot to the wilderness for parties seeking to fish and hunt in unsullied lands. Over dinner Adams talked of flying and crosswinds and groundloops, asked some questions, answered some. He invited the cold, oily barnstormer to his home, to meet his family, to look at photographs of airplanes and flights gone by.

At five-thirty the next morning he was down to take the aeronaut to breakfast, and to help him start the engine. Another takeoff, a wing-rocking farewell, and long cold morning hours in the twisting knife-wind as the sun pulled itself into the sky.

We followed U.S. Highway 80 for several hundred miles through the wilderness of western Texas, most of it at a five-foot altitude above the deserted road to avoid an ever-present headwind. The big land is always there, always waiting, always watching every turn of the propeller of airplanes that dare cross it. I thought of my survival rations and water jug, and was glad they were along.

Ahead, a thunderstorm, standing upon its wide slanting pillar of hard gray rain. "An adventure waits!" I said to the Parks, and pulled the seat belt tighter. I could follow the railroad to the right and avoid the rain, or the road to the left and fly through it. I've always thought it a good practice to pick up gauntlets when they're thrown, so we followed the highway. Just as I had completed tying myself to the mast, as it were, and the first drops of rain slashed across the windscreen, the engine stopped. One adventure at a time, I quickly began thinking, and as we wheeled hard to the right, I was thinking of the survival kit. The desert looked terribly empty. On her own, the Whirlwind gasped back into life, sputtering and choking. Fuel was on, mixture rich, plenty of fuel in the tank. The magnetos. The magnetos were wet. Switch to right mag and the Whirlwind ceased her coughing and purred along, blinking her eyes. Switch to left mag and she stopped cold, misfired, backfired. Switch quickly back to right mag. Map, map, where's the map? Nearest town is, let's see . . . (wind roar increasing in the wires) . . . is Fabens, Texas, and twenty miles west: between here and Fabens . . . (wind screaming now) . . . Oh, not now, airplane. I'm just looking at the map! Isn't that all right? Pull the nose back up to the horizon, move the stabilizer trim up a notch . . . Fabens is twenty miles and if I follow the railroad it will turn to the left . . . (wind dying away, going soft and quiet, shadows shifting across the map) . . . OK! OK! Please don't give me a hard time here. Can't you see that desert down below? Do you want to lose a wing or a wheel on one of those rocks?

The Parks settled down to follow the railroad, but whenever I wanted to scare myself, I twisted the magneto switch to LEFT and listened to the engine choke and die away. It was a comfort, minutes later, to land in the blowing sand of Fabens, Texas. I spread a sleeping bag under the wing, with parachute and jacket for a pillow, and dreamed no dreams.

In the morning the magnetos were dry and ready for business, and business was seven hundred miles of desert. Our country certainly does have a lot of sand in it. And rocks. And weeds growing brown in the sun. And railroad tracks straight as fallen pines stretching away to the horizon.

As we were crossing the border into Arizona, the left mag began complaining again. So it was five hundred miles on the right mag, between the gunnery ranges south of Phoenix, through the dust storm over Yuma. It got so that the left magneto didn't scare me at all. So that one magneto

can run the engine if the other one quits. Airplanes used to have single-ignition engines. If the right magneto fails, I land on U.S. Highway 80 and break out the survival kit. By Palm Springs, California, the left mag was working again. Must be when it gets hot it quits; let it cool for a while and it's OK.

Almost home, I thought. "Almost home," I said to the Parks. "Won't be long now."

But there were storms west of the mountains, and rain, and great winds swept down the passes. If only I had the Fairchild with its instruments and radios! We tried the pass at Julian, the Parks and I, and were shaken and whipped and thrown back into the desert for our audacity. We tried to pass to San Diego, and for the first time in my life, indicating seventy-five miles per hour, I was flying backward. An eerie feeling, one that makes one look quickly to the airspeed indicator for assurance. But assurance notwithstanding, the Parks was simply incapable of flying west against the wind. Then north again, to a long and personal battle with the pass at Banning, and with Mount San Jacinto. You big bully! I thought, and glared up at the mountain, its peak swirling in storm cloud and snow. We tried the rain again, and this time the magnetos, angry at the mountains, didn't mind it at all.

Still it was fight and fight and fight till we finally clawed our way to land on the rain-slick runway at Banning.

An hour later, rested and ready for more fighting, I saw a break in the clouds to the west, over a low range of hills. We took off and caught the rain again, rain that stings like steel pellets thrown and rain that washes one's goggles bright and clean. And with it, turbulence from the wind over the hills so that the engine stopped time and again as negative G pulled fuel from the carburetor.

And suddenly it was all over. The last range of hills was

past, and ahead were clouds broken with giant shafts of sunlight streaking down. Suddenly, like flying over into the Promised Land, as though a decision had been made that the little Parks had fought hard enough, had proved herself, and now the fight would not be necessary. One of those moments a pilot doesn't forget: after the gray whistling steel-shot rain, sunlight; after the smashing turbulence, mirror-smooth air; after glowering mountains and furious cloud, a little airport, a last landing, and home.

Miss that Big Briefing In The Sky, and you have to find out for yourself about flying coast to coast in old airplanes. If you don't get the word from someone else, an airplane has to teach you.

And the lesson? People can fly old open-cockpit biplanes thousands of miles, can learn things of their country, of the early pilots to whom aviation owes its life, and of themselves. Something perhaps, that no Briefing could ever teach.

There's always the sky

I was supposed to write a story about the man, not to kill him in cold blood. But somehow I couldn't make him believe that—it was one of those rare times that I had met a person so frightened he was alien, and I stood helpless to talk with him as though I spoke ancient Urdu. It was disconcerting, to find that words sometimes have no meaning, and no effect at all. The man who was to have been the central figure of the story advised me clearly that he was on to me, he knew that I was a puppet, a boor, an ingrate, and a mob of other unsavory characters all wrapped in a faded flying jacket.

A few years earlier, I might have experimented with violence to communicate with him, but this time I chose to leave the room. I walked out into the night air, and in the dim moonlight by the shore of the sea—for this was to have been a story of the man and his resort paradise.

The breakers boomed along the dark beach, flickering blue-green-phosphorous like gentle peaceful howitzers firing in the dark, and I watched the salt ocean rush in swift and steady, slow and back, hissing softly. I walked half an hour perhaps, trying to understand the man and his fear, and finally gave it up as a bad job. It was only then, turning away from the ground, that I happened to look up.

And there, over the elegant resort lands and over the sea, over the oblivious guests at the indoor bar and over me and all my little problems, was the sky.

I slowed, there on the sand, and at last stopped and looked way on up into the air. From past-horizon north to past-horizon south, from beyond land's end to beyond the depths of the western sea, lived the billion-mile sky. It was very calm, very still.

Some high cirrus drifted along under a slice of moon, borne ever so carefully on a faint, faint wind. And I noticed something that night that I had never noticed before.

That the sky is always moving, but it's never gone.

That no matter what, the sky is always with us.

And that the sky cannot be bothered. My problems, to the sky, did not exist, never had existed, never would exist.

The sky does not misunderstand.

The sky does not judge.

The sky, very simply, is.

It is, whether we wish to see that fact or to bury ourselves under a thousand miles of earth, or even deeper still, under the impenetrable roof of unthinking routine.

It happened a year later that for some reason I was in New York City, and everything was going wrong and my total assets equaled twenty-six cents and I was hungry and the very last place I wanted to be was in the prison streets of sundown Manhattan, with iron-barred windows and quintuple-lock doors. But it happened that I looked up, which is something one never does in Manhattan, of course, and again, as it had been by the sea—way on up there, way high over the canyons of Madison Avenue and Lexington and Park—was the sky. It was there. Unhurried. Unchanged. Warm and welcome as home.

What do you know, I thought. What do you know about that. No matter how tangled and twisted and distressful goes the life of an airplane pilot, he always has a home, waiting. For him always waits the joy of being back in the air, of looking down at and up to the clouds, for him always waits that inner cry, "I'm home again!"

"Bunch of mist, bunch of empty air," the people of the ground will say. "Get your head out of the clouds, get your feet on the ground." Yet in times as far separated as that lonely beach and the crowded Manhattan street, I was lifted from black despair into freedom. From annoyance and anger and fear to a thought, Hey! I don't care! I'm happy!

Just by looking into the sky.

This kind of thing happens, perhaps, because pilots aren't far-traveling wanderers after all. It may be that pilots are happy only when they are at home. And it may be that they are home only when they can somehow touch the sky.

Steel, aluminum, nuts and bolts

An airplane is a machine. It is not possible for it to be alive. Nor is it possible for it to wish or to hope or to hate or to love.

The machine that is called "airplane" is made of two sections, the "engine" and the "airframe," each of which is built of common machine-building materials. There is no secret, no dark magic, there are no incantations said over any airplane in order to make it fly. It flies because of known and invariable laws which cannot be changed for any reason.

An "engine," briefly, is a block of metal that has been drilled with certain holes and set with certain springs and valves and gears. It does not in any way come into life when it is bolted on the front of an airframe. Those vibrations through an engine are caused by the rapid burning of fuel within its cylinders, by the action of its moving parts, by the forces that a spinning propeller creates.

An "airframe" is a sort of cage built of steel tubing and sheet aluminum. It is tin and fabric and wire. It is nuts and bolts. An airframe is made to the calculations of the aircraft designer, who is a very wise and practical man who makes his living at this sort of thing and does not mess around with esoteric mumbo-jumbo.

There is no part in any airplane for which there does not exist a blueprint. There is no part which cannot be unscrewed into simple plates and castings and forgings. The airplane was invented. It did not "come into being," it was never brought to life. An airplane is a machine as an automobile is a machine, as a chain saw is a machine, as a drill press is a machine.

Is there a voice in reply to this, from perhaps the newest of student pilots, saying that an airplane is a creature of the air, and so has special forces acting upon it that a drill press does not have?

Wrong. An airplane is not a creature. It is a machine: blind, dumb, cold, dead. Every force upon it is a known force. A million hours of research and flight tests have shown us all there is to know about an airplane: Lift-Weight-Thrust-Drag. Angles of attack, centers of pressure, power required versus power available, and parasite drag increases as the square of the airspeed.

Yet there are a few airplane pilots who somehow want to believe that this machine is an animal, that it is alive. Make certain that you do not believe it. That is absolutely impossible.

The takeoff performance of any aircraft, for instance, depends upon wing loading, power loading, airfoil coefficients, and upon density altitude, wind, slope and surface of the runway. All these are things that can be measured with tape measures and test machines, and when they are run through charts and computers, they give us an absolute minimum takeoff distance.

There is no sentence, no word, no hint in any technical manual ever printed that even remotely says that this machine's performance can possibly change because of a pilot's hopes or his dreams, or his kindness to his airplane. This is critically important for you to know.

I give you an example. I give you a pilot. Let's say that his name is . . . oh . . . Everett Donnelly. Let's say that he learned to fly in a 7AC Aeronca Champion. N2758E.

Then later, let's say that Everett Donnelly became a first officer with United Air Lines, and then a captain, and that for fun he began looking for that same old Aeronca Champ. Let's say he asked questions and wrote letters and searched for a year and a half across the country, and that at last he found what was left of N2758E, smashed under a fallen hangar at an abandoned airport. Let's say he spent just under two years rebuilding the airplane, touching and finishing every nut and bolt and pulley and seam of it. And then perhaps he flew that Champ for five years, and perhaps he refused quite a few good offers from people who wanted to buy it, and perhaps he kept it in perfect condition because it was a part of his life that he enjoyed and because the airplane itself had become something that he loved.

Now let's say that one day he landed in a high mountain field with a broken oil line. Let's say he fixed the line, added oil from cans he always carried, and was ready for takeoff.

Now read this next part carefully. Let's say that if Everett Donnelly does not take off at this time, he will be buried in the blizzard of December 8, 1966. Let's say that there is no road to this mountain field, no civilization nearby. And let's say that there is a stand of sixty-foot pines all around the field and that there is no wind.

I give you this situation. I then set these figures into a computer that is programmed with this particular Champ's performance specifications and with this particular field's terrain and atmosphere. The final sum that the computer presents, after clicking for a while, is a minimum distance of 1594 feet to clear a sixty-foot obstacle, assuming perfect pilot technique.

Everett Donnelly, not knowing as precisely as a computer, but knowing that the takeoff will not be an easy one, paces the distance at 1180 feet from the start of the roll to the base of the trees facing him. By pulling the tail of his machine back between two trees, he can increase his field length to 1187 feet. This means nothing. The field is 407 feet too short.

And now I give you some facts that cannot possibly make any difference in the takeoff roll of Aeronca Champion N2758E.

Let's say that Everett Donnelly thinks of the blizzard on its way, of his cold death and the destruction of his airplane if he does not fly out of this field at once.

He remembers the first day that he saw the Champ, sun-yellow and faded red-earth trim, splashed with mud, hopping passengers and flying students from a little field in Pennsylvania after the war. He remembers working weekends and all summer to pay for learning to fly this machine.

He remembers fifteen thousand flying hours, and of finding the Champ again, under the hangar.

He remembers the years rebuilding and of Jeanne Donnelly's first flight in it and that she will fly in no other machine than N2758E.

He thinks of his son's first flight and instruction, and of his solo only a week earlier, on the morning of the boy's sixteenth birthday.

And he swings the propeller of his machine and he steps into the cockpit and pushes the throttle all the way forward and the Champ begins to move toward the trees at the other end of the field, because it is time to go home.

Please believe that my research about airplanes is complete. There are no flaws in it. That research covers all the learning of all the aeronautical engineers and aircraft designers and mechanics since man first began to fly. There is no theory that these people have not checked and proved in practice.

And every one of them and every one of the facts are dead set against there being any hope for Everett Donnelly if he tries to make that takeoff from a field that is 407 feet too short. Better dig a cave and try to survive the blizzard; better let the airplane blow to shreds in the wind while the pilot tries to walk out of the mountains; better anything than try to clear an obstacle that is absolutely impossible to clear.

An airplane, I have shown, is a machine. This is not my idea, it is not what I have wished into being. It is not even me, writing this, but the tens of thousands of brilliant minds that have given mankind the speed and the technology of flight. All that I have done is merely ask, in my research, if any of them believes that an airplane is anything more than a machine. And in a thousand books and a half-million pages and diagrams and formulas, there is not one word, there is not one unspoken hope set against the mathematics and the computation of Everett Donnelly's takeoff

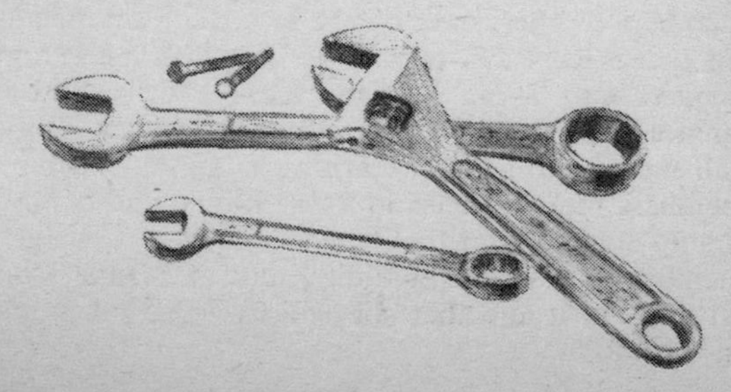

roll. Not one voice said that if conditions are right, that if a pilot loves his airplane and shows this in his care, then an airplane might just one time and for the briefest of moments become a thing alive, that can love in return and show this in its flight. There was not one word that said this could be so.

The computer clicked its answer and that was final. The figure given was the absolute minimum takeoff distance: 1594 feet.

There was no error, I assure you. The Champ could not possibly clear those trees. It was impossible for it to do so. By precise calculation, it must hit the trees twenty-eight feet above ground level at a true airspeed of fifty-one miles per hour. The impact, centered upon the right main wing spar, seventy-two inches from the wing-fuselage attach fittings, would be of sufficient force to collapse the main and rear spars. The inertia of the remaining aircraft weight, acting through a new center of gravity, would whip the aircraft to the right and toward the ground. Impact with the ground would cause stress on the engine mount in excess of design load factors. The engine would move backward through the firewall and fuel tank. Gasoline sprayed across the exhaust manifold would make a flammable vapor that would be ignited by exhaust flame from the broken cylinders. The basic structure of the aircraft would be consumed by fire in four minutes thirty-seven seconds, which may or may not be sufficient time for the occupant to recover from any induced unconsciousness of the impact and leave the machine. The last point, the sufficiency of that time, is uncertain because it does not fall within the realm of aerodynamics and stress analysis.

The whole point of my report to you, then, is that you remember this: The airplane that you fly is a machine. If you love it and treat it well, it is a machine. An airplane is a machine.

It is not possible for me to have seen Everett Donnelly this morning, shooting landings in his Champ and taxiing in for gas.

I couldn't have said, "Everett, you're dead!"

He couldn't have laughed at me. "You gone crazy? I'm no more dead than you are. Tell me, how did I die?"

"You went down in the mountains, forty-two miles north of Barton's Flat and the field was only 1187 feet long and

N2759E

the density altitude was 4530 feet and your wing loading was 6.45 pounds per square foot."

"Oh, that. Sure I was down. Oil line broke. But I put a hose clamp on it and added some oil and took off again and flew home before the storm. Couldn't very well stay there, could I?"

"But the takeoff roll . . ."

"You better believe it! I had pine needles in the landing gear when I got home. But the old Champ will do nice things every once in a while, if I'm good to her."

It is impossible for that to have happened. It is impossible for anything like that ever to have happened. If you have ever heard of anything like this ever happening to any pilot, if anything like this has ever happened to you, it could not have happened. That would be impossible.

An airplane cannot live.

An airplane cannot possibly know what "love" is.

An airplane is cold metal.

An airplane is a machine.

The girl from a long time ago

"I want to go with you."
"It's going to be cold."
"I still want to go with you."
"And windy and oily and so loud you won't be able to think."
"I know, I'll wish I had never done it. But I want to go with you."
"And sleeping under the wing at night and storms and rain and mud. And little small cafes in little small towns is where you'd eat."
"I know."
"And no complaining allowed. Not one complaint."
"I promise."

And so, after teetering on her own quiet brink for days I could not number, my wife told me she wanted to ride the wild roaring front cockpit of my 1929 barnstorming biplane, on a flight planned to cross thirty-five hundred miles of spiky western America—across the Great Plains into the low hills of Iowa, and back to California through the Rocky Mountains and the Sierra Nevada.

I had a reason to make the flight. Once a year a thousand clattery slow harp-wire flying machines, antiques out of old skies, converge for a week at a grasscarpet airfield in the middle of summer Iowa. It is a place where pilots talk of dope-and-fabric joys and oil-sprayed sorrows, each glad for friends as mad and as loving of aeroplanes as himself. They are a family, these people, and I was one of them; the reunion was to be, and that was all the reason I needed to go there.

It was harder for Bette. She had to admit, as she arranged for two weeks of child-care, that she was making the flight because she wanted to go, because it would be fun, because she could say that she had done it. That took courage, of course, but I wouldn't help but wonder whether she could make it, and I was convinced that she hadn't the first idea of what that flight was going to be like.

I had made one long flight in the biplane, bringing it home to Los Angeles from North Carolina, a week after I bought it from an antique airplane collector. During that flight I had one minor crash, one engine failure, three days of freezing cold, and two days over the desert that were so hot the engine temperatures rose to their limits. I had fought winds that pushed the biplane backward, and one time had to fly so low under clouds that my wheels were brushing the treetops. I had more than enough to worry about on that flight, all by myself, and this one, with my wife, was to be a thousand miles longer.

"You're sure you want to do this?" I asked as I rolled the biplane from the hangar, the sun lifting its first faint dawnlight into the sky. She was intently rummaging under our sleeping bags, adding one last item to the survival kit.

"I'm sure," she said absently.

I have to admit that I held a certain savage curiosity to see how she would handle the adventure. Neither of us have much interest in camping or roughing it; we like to read, to see a play now and then, and, because I was a pilot in the Air Force, we like to fly. I enjoy my airplane, but I have a great deal of respect for it. Only the day before, I had finished repairing the engine after its fifth failure in as many months. By now, I hoped, it had all its troubles repaired out of it, but nevertheless I vowed to fly so that I could always glide to some kind of level ground if the engine failed again. I was not taking bets on whether we would make it to Iowa at all—the odds were about fifty-fifty.

None of this turned her head.

Now, I thought, as I cranked the old engine into its deafening blasting blue-smoke life, as I checked its instruments and let it warm, I'll find out just what kind of wife I married, seven years ago. For Bette, strapped in the open cockpit, dressed in a 1929 flying costume beneath a huge furry coat, lashed already by propeller blast, the test had begun.

An hour and a half later, at a temperature of twenty-eight degrees, we were joined in flight by two other antiques, both of them closed-cabin monoplanes, both, I knew, with heaters. Cruising at five thousand feet and ninety miles per hour, I moved closer to my friends' airplanes, and waved. I was glad to have them there. If our engine quit, we wouldn't be alone.

Flying within a few yards of the monoplanes, I could see the wives were dressed in skirts and blouses. I shivered under my scarf and leather jacket and wondered in that early-morning air if Bette was already sorry for her decision.

Though our two cockpits were only three feet apart, the wind and the engine roared so furiously about us that even a shout couldn't be heard. We carried no radio, no intercommunication system. Whenever we had to talk to each other, it was in sign language, or by passing a wind-battered scrap of paper with words scrawled in bouncing letters.

In that moment that I was shivering and wondering if my sheltered wife was about ready to admit that this was all a foolish mistake, I saw her reach for her pencil. Here it comes, I thought, and I tried to guess how she would word it. Would she write "Let's quit," just like that? Or "Can't stand the cold"? Our breath came in white frost-puffs swept instantly overboard. Or just "Sorry"? Depends on how cold and wind-blasted she is. I could see the spray of engine rocker-box grease across her windshield and I saw it on her goggles as she turned to hand me the note, her tiny, thin-gloved fingers extending from the huge furry sleeve. Holding the airplane control stick between my knees, I reached for the scrap of folded paper. We were only one hundred fifty miles from home, and I could fly her back in two hours.

There was one word written. "FUN!" With a little laughing face drawn alongside.

She was watching me read, and when I looked up, she smiled.

What can you do, with a wife like that? I smiled back, touched my glove to my leather helmet, and saluted.

Three hours later, after a brief stop for fuel, we were over the heart of the Arizona desert. It was almost noon, and even at five thousand feet the wind was hot. Bette's coat was piled on the seat beside her, the top of it whipping in the heated propeller blast. A mile below, and as far as we could see, the meaning of "desert." Barren piles of jagged rock, mile on mile of sand, utterly and completely empty.

Once again I was glad for our companions. If the engine chose this moment to fail, it would be simple to come down on the sand, not even damaging the airplane. But it was blazing, rippling hot down there, and I was grateful in the thought of the water jug that we had packed into our survival kit.

Then it struck me full force, in delayed action. By what right did I even consider allowing my wife in that front cockpit? If the engine stopped, she would be five hundred miles from her home and children, standing by one tiny speck of a biplane in the center of the biggest desert in America. With sand and snakes and a scorching white sun and not a blade of grass or shred of tree as far as she could see. What kind of blind, unthinking, irresponsible husband was I, to allow that girl, my own wife, to be exposed to this? As I stormed at myself, Bette looked back at me, and gave her hand signal for "mountain," all her gloved fingers together and pointing upward. Then she scowled dark, over the top of her signal to show that this was an especially mean mountain, and pointed down.

She was right. But the mountain was only a fraction meaner than all the rest of the dead land about us.

In seeing the land, though, I found my right to bring her there. In her mountain signal, the wife that I had tried so hard to shelter and protect was discovering her country, seeing it as it was. As long as she could see it this way, and with joy instead of fear, with gratitude instead of concern, that was my right to bring her. In that moment, I was glad that she had come.

Arizona rolled by, and the desert gave grudging way, an inch at a time, to higher land and scrub pine. Then in a rush it surrendered to broad forests of pine, and tiny rivers, and some lonely pastures, with far-set ranch houses.

The biplane rolled smoothly through the sky, but I was concerned. The engine oil pressure was not behaving properly. It slowly fell back from sixty pounds pressure to forty-seven. This was still within its limits, but it was not right, for oil pressure in an airplane engine should be a very steady thing.

Bette was asleep now in the front cockpit, letting the wind sweep over her head as she rested on a mound of furry coat. I was glad she slept, and concentrated on mental diagrams of the inside of the old engine, trying to think of what the trouble could be. Then, two thousand feet above the ground, the engine stopped. The silence was so unnatural that Bette awoke, and looked down for the airport where we must be landing.

There was none. We were fifty miles from any airport, and the more I worked over the engine, moving fuel selectors, setting ignition switches, the more I knew that we would never make it to an airport.

The biplane sank swiftly out of the sky, and I rocked the wings to our friends, telling them that we were having a little trouble. They turned toward us immediately, but they could do nothing more than watch us go down.

Forests carpeted the mountains behind and the mountains ahead. We were gliding down into a narrow valley, and along the edge of that valley, a ranch, and a fenced pasture. I turned toward the pasture. It was the only strip of level ground as far as I could see.

Bette looked back at me, and raised her eyebrows. She didn't seem frightened. I nodded to her that everything was all right, and that we were going to land in the pasture. I was ready to allow her to be frightened, for I would have been had I been in her place. This was her first forced landing; it was my sixth. One part of me stopped to watch her critically, to see how she took this engine failure—this event that, as far as the newspapers told her, inevitably resulted in a gigantic fatal crash and tall black headlines.

There were two fields, side by side. I chose the one that looked the smoothest, making one final gliding circle to land. Bette pointed near the other field, raising her eyebrows in question. I shook my head no. Whatever you are asking, Bette, no. Just let me land the airplane now, and we'll talk later.

The biplane swept down, losing height quickly, crossed the fence, and slammed hard onto the ground. It bounced one time back into the air, then came down again, bumping and thundering across the rough hard field. I hoped there weren't any hidden cows. There were some on the hillside. In a few seconds' rolling, my question about the cows became an academic thought, for we were down, and stopped. It was utterly quiet, and I waited for my wife's first comment after her first forced landing. I tried to guess what she would say. "So much for Iowa"? "Where's the nearest rail-

road"? "What are we going to do now"? I waited.

She lifted her goggles and smiled.

"Didn't you see the airport?"

"WHAT!"

"The airport, dear. A little field over there, didn't you see it? It has a windsock and everything." She hopped down from her cockpit and pointed. "See?"

There was a windsock, all right. The only minor balm I held was that the single dirt runway looked shorter and rougher than the pasture we had landed upon.

That part of me that was watching and checking and grading my wife, and that was all of me, at that moment, broke down and laughed out loud. Here was a girl I had never met, I had never seen before. A beautiful young lady with tousled hair and engine oil edging a big white goggle-print around her eyes, smiling impishly up at me. I have never been so helplessly charmed as I was on that after-noon by this incredible young woman.

There was no way to tell her how well she had passed her test. The test was over and done in that moment, and the book thrown away.

For a second the ground shook as our companions roared low overhead. We waved that we were all right, and pointed that the biplane was undamaged. They dropped a message, saying that if we signaled, they would come down and land. I waved them away. We were in good shape. I had some antique-airplane friends in Phoenix who would be able to help with the engine. The monoplanes flew low one more time, rocking their wings, and disappeared over the mountains to the east.

That night, after the engine was repaired, I said hello to the lovely young woman who rode the front cockpit of my airplane. We unrolled our sleeping bags in the icy dark, heads together, and looked out at the whirling blazing center of our galaxy, and talked about what it felt like to be a creature that lived along the edge of so many suns.

My biplane had carried me back into its own year, into 1929, and these hills around were 1929 hills, and those suns. I knew what a time-traveler felt like, to drift back into the years before he was born, and there to fall in love with a slim dark-eyed young mistress in flying helmet and goggles. I knew that I would never return to my own time. We slept, that night, this strange young woman and I, on the edge of our galaxy.

The biplane thundered on across Arizona and into New Mexico, without the monoplanes at its side. Long hard flights it made; four hours in the cockpit, a moment out for a sandwich and a tank of fuel, a quart of oil, and back again into the wind. The windswept notes that my new wife handed back showed a mind as keen and bright as her body. They reflected a girl looking on a new world, with eyes bright for seeing.

"The red balloon-sun bounces up from the horizon at dawn as if a child has let go of its string."

"Pasture sprinklers in early morning are white feathers evenly strung."

These were the sights I had seen in ten years of flying, and had never seen, until someone else who had never seen them either, framed them on scraps of notepaper and passed them back to me.

"The free-form ranches of New Mexico give way only gradually to the precise checkerboard pattern of Kansas. The top of Texas passes by underwing incognito. Not even a fanfare or oil well to mark it."

"Corn from horizon to horizon. How can the world eat so much corn? Corn flakes, corn bread, corn muffins, corn on the cob, corn off the cob, cream corn, corn puddin, corn meal mush, corncorncorn."

Now and then, as we flew, a utilitarian question. "Why are we headed for the only cloud in the sky, answer me that!" Question answered with a shrug, she went back seeing and thinking.

"Kind of takes the fun out of passing a train when you

can see the engine and caboose at the same time."

A prairie city moved majestically toward us, steaming in from the ocean of the horizon. "What city?" she wrote.

I mouthed the name.

"HOMINY?" she wrote, and held up the paper in front of my windshield. I shook my head and mouthed it again.

"HOMLICK?"

I said it over and over again, the word whipped away in the slipstream.

"AMANDY?"

"ALMONDIC?"

"ALBANY?"

"ABANY?"

I kept saying the name, over and over, faster and faster.

"ABILENE!"

I nodded, and she peered down over the side of her cockpit at the city, able now to properly inspect it.

The biplane flew three days into the east, content to have brought me back to its time and introduced me to this quick young person. The engine didn't stop again, or falter, even when cold rain poured down upon it, on the last miles into Iowa.

"Are we escorting this storm to Ottumwa?"

I could only nod and wipe the spray from my goggles.

At the fly-in, I met friends from around the country, my wife quiet and happy at my side. She said little, but listened carefully, missing nothing with her bright eyes. She seemed glad to let the wind play in her midnight hair.

Five days later, we struck for home again, I with the hidden fear that I must return to a wife that I no longer knew. How much I would rather stay and roam the country with this mistress-wife!

"A fly-in," her first note read, hours out of Iowa and over the plains of Nebraska, "is individual people; where they've gone, what they've done, what they've learned, their plans for the future."

And then she was quiet for a long time, looking out upon the two other biplanes with whom we returned west, the three of us flying together into great flaming sunsets every evening.

The hour came, as I knew it must, when we had crossed plains and mountains and desert once again, leaving them holding their challenge thrust silently into the sky. Her last note read, "I think America would be a happier place if every citizen, on reaching the age of eighteen, would be given an aerial tour of the entire country."

The other biplanes waved goodbye, and banked in steep sudden turns away from us, toward their own airports. We were home.

Biplane once again back in its hangar, we drove quietly to our house. I was sad, as I am sad when I close a book and must say goodbye to a heroine that I have come to love. Whether she is real or not, I wish that I could spend more time with her.

She sat beside me in the car as I drove, but in a few minutes it would be all over. She would comb her midnight hair neatly into place, away from the wind and the propeller blast, to become once again the focus of her children's demands. She would walk again back into the world of shelter, a routine world that does not ask her to see with bright eyes, or to look down upon desert mountains, or to fight lofty windstorms. A routine that has never seen a double or full-circle rainbow.

But the book was not quite closed. Sparkling now and then, here and there, at strange and unexpected times, the young woman that I discovered in 1929 and that I loved before I was born, looks up at me impishly, and there is the faintest hint of engine oil around the eyes. And she is gone before I can speak, before I can catch her hand and tell her wait.

Adrift at Kennedy airport

When I first saw Kennedy International Airport, there was no doubt that it was a place, a great island of concrete and sand and glass and paint, and derricks tilting their steel necks and taking rafters in their teeth and lifting them through the air to new constructions, to alloy roof-trees in a burnt-kerosene sky. It never occurred to me to doubt that. It was a sterile dark desert before dawn, it was pandemonium and a vision of the next century's rush hour, when the jets lined up forty or sixty in rows waiting for takeoff, and arriving flights landed five hours late and children sat on baggage and cried and once in a while a grownup cried too.

But the longer I watched, the more I began to see the fact: Kennedy isn't a place quite as much as it is a cement-and-iron thought, with solid sharp edges at the corners; a proud stone idea that we have some kind of control over space and over time, and here within these boundaries we have all decided to get together and believe it.

Somewhere else is abstract wonder about shrinking worlds and five hours to England and lunch-in-New-York-dinner-in-Los-Angeles. But here there is no abstract, there are no vague discussions. Here it happens. At ten o'clock on our watch we walk aboard BOAC Flight 157 and we expect, by three o'clock, either to have been killed in a monster crash or to be hailing a taxi in London.

Everything at Kennedy has been built to make that idea fact. The concrete is there for that cause, and the steel and the glass, the airplanes, the sound of engines; the ground itself was trucked in and poured over Jamaica swamps to make that idea fact. No lectures here about cutting space-time into shreds, here's where we do it. We do it with the sweeping blur of a wing in the air, with that ground-rumbling full-throttle blast of mammoth engines leaning hungry into the wind, round metal mouths open as wide as they'll go, devouring ten tons of air a minute, attacking it cold, torching it with rings of fire till it's black with agony, blasting it a hundred times faster out carbon tailpipes, turning empty air to heat to thrust to speed to flight.

Kennedy Airport is a fine act, by an excellent magician. No matter what we believe, London will appear in five hours before our eyes, and, finishing lunch, we'll have dinner in Los Angeles.

Crowds. I don't like crowds. But why, then, do I stand here at the rush hour in one of the biggest airports in the world, and watch the thousands of people swirling about me, and find myself happy and warm?

Perhaps it is because this is a different kind of crowd.

The rivers of people anywhere else in the world, pouring along sidewalks, pressing through subways and train stations and bus terminals in the morning and evening, are rivers of people who know just where they are and just where they are going, they have passed this way before and they know that they will pass this way again. So knowing, not much humanity shows in the masks they wear—that humanity lies within, struggling with problems, contemplating joys of past and future. Those crowds aren't people at all, but carriers of people, vehicles with people inside, all shades drawn. There is not much to be said for watching a procession of curtained carriages.

The crowds at Kennedy Airport, though, do not come this way every morning and every evening, and no one is quite sure of just where he is or just where he should be. With this, a misty state of emergency invests the air, in which it is all right to talk to a stranger, to ask for directions and help, all right to lend a hand to somebody a little loster than we. The masks are not quite so firmly in place, the curtains not quite so fully drawn, and you can see the people inside.

It occurred to me, standing on the second-floor balcony, looking down, that these are the people from all over the world who are making their nations run, these are the ones directing the path of history. It was startling, the intelligence to be seen in that humanity, and the humor, and the respect there for others. These are the people in control of the governments, the ones who protest wrongs, and change them; these are the members of the final jury of their land, with more power than any court or military, who can overthrow any injustice that reaches their combined hearts, these are they whose ideals are appealed to by men who seek the accomplishment of any good thing. For these, newspapers are printed, things are created, films are made, books are written.

There must be criminals in the crowds at Kennedy, too, there must be petty small men, and greedy and cruel. But they must be greatly outnumbered, else why that warmth I know, watching them all?

Here in the currents of the International Arrivals Building, for instance, is a dark-haired girl in wine-colored traveling clothes, moving slowly through a packed crowd that she wishes to move swiftly through. It is eight-fourteen of a Friday evening. She works her way toward the automatic doors at the north wall of the building, perhaps arriving, perhaps leaving. Her face is not quite set, she is paying some attention to the problem of moving, but not a great deal; she is patiently forging ahead.

Now from her right the crowd has given way to a heavy steel baggage cart, a moving hillock of leather and plaid. She does not notice it coming, bearing down upon her. It is her turn to give way to the cart, and still she does not see it as she moves toward the door.

"LOOK OUT, PLEASE!" The porter shouts and tries to brake the cart in the last instant before it gently rams her. He does turn it slightly, and the iron wheels roll two inches in front of her.

The dark-haired girl in the wine-colored suit sees the cart at last, stops instantly, in mid-step, and without making a sound she grimaces "EEEK!"

The cart rumbles past as she smiles at herself for her drama, at the porter in apology for not watching out.

He says a word, "You be careful, miss," and they go their way, smiling still. She is gone out one door, he out another, and I stand there and watch and somehow feel tender and loving toward all mankind.

It is like watching a fire, or the sea, this watching of people at Kennedy, and I stood quietly there for weeks, munching a sandwich sometimes, just watching. Meeting, knowing, bidding farewell in the course of seconds to tens of thousands of fellow people who neither knew nor cared that I saw, going their way about the business of running their lives and their nations.

I don't like crowds, but some crowds I like.

> The form said:
> *Lenora Edwards, age nine. Speaks English, minor traveling alone; small for her age. Address Martinsyde Road Kings Standing 3B Birmingham, England. She is arriving alone on TWA and is making a flight to Dayton, Ohio. Please meet and assist with transfer. Child is coming for three-week visit with her father. Parents divorced.*

For one day I joined Traveler's Aid, because I've always been curious about Traveler's Aid, seen them at their little posts at train stations and never really aiding anybody, that I could see.

Marlene Feldman, a pretty girl, former legal secretary, was the one who took the form, handed me a Traveler's Aid armband, and led the way to the International Arrivals Building. Our little girl's flight was to have arrived at three-forty on a holiday weekend. At six o'clock we learned that by seven o'clock we might know what time it would be expected to land.

"She will probably not make her connection," Marlene said in a voice that was used to preparing for the worst. She must have been a good legal secretary. Now she was unruffled and in control, grasping the threads of unraveled plans and trying to weave them back together, for the sake of Lenora Edwards.

"You can be around this every day, but every time you see a plane take off or land, it's still fascinating. It's just beautiful. And every time you see one go up, you say, 'I wish I was on it . . .' Hello, United? This is Traveler's Aid, and we need a late flight from Kennedy to Dayton, Ohio . . ."

There was no late flight to Dayton.

By eight p.m. the flight with Lenora Edwards on board still had not landed, the airport was a choking swarming mass of passengers and passengers' friends come to meet them, the sound of engines in the air.

Marlene Feldman, telephone in hand, was supposed to have finished her working at five p.m., it was now eight-thirty, she had had no dinner.

"Just a minute. One more call and we'll go eat." She dialed TWA for the twelfth time, and at last they had an expected arrival . . . Lenora's flight would be unloading in twenty minutes.

"Well, there goes dinner," Marlene said. Which wasn't quite the truth. The restaurants at Kennedy were crowded, even the lines waiting at them were crowded, but the candy machines were almost unpatronized. She had a Sunshine Peanut Cheese Sandwich Lunchie for her dinner, I had a Hershey bar.

We found Lenora in the crowd by the Customs area, waiting for her one piece of baggage, a white suitcase.

"Welcome to America," I said. She didn't reply.

She did talk to Marlene, in a very clear little British voice, "I suppose I've missed my plane, haven't I?"

"I'm afraid so, honey, and there's not another flight going out till tomorrow morning. But don't worry. We'll get it all fixed up for you. Did you have a nice flight coming over?"

We breezed through Customs without even stopping at the desk, and I hoped faintly that the white suitcase I carried wasn't packed with diamonds or heroin. It didn't feel like it, but those things are hard to tell.

The crowd by now was a New Year's Times Square crowd, and we wedged slowly through it to the office. Excuse us. Excuse us, please. Could we get through? What was the poor little girl thinking? All this chaos, met by two strangers, missed her flight, no plane out till tomorrow? She was calm as a teacup. If I were nine years old in that place, five hours late in a foreign country, I would have gone up in kind of green smoke.

Marlene was on the telephone again, calling the girl's father, collect to Dayton. "Mister Edwards. Traveler's Aid, Kennedy Airport. We have Lenora here, she missed the flight to Dayton, so do not go to the airport. She'll stay here tonight, we'll arrange for that. I'll call you back just as soon as I know what's happening."

"How are you doing, honey?" she said, dialing again on the phone.

"Just fine."

It was arranged. Lenora would stay that evening at the International Hotel with a TWA stewardess from the flight on which she arrived, who would bring her to the United Air Lines Terminal in the morning.

The telephone again to the father, to give him the name and number of the stewardess and the hotel. "Lenora will be arriving Flight 521, into Dayton at ten twenty-six in the morning. That's right. Yes. Yes. Of course I will," Marlene said. "You're quite welcome."

"OK, Lenora," she said when the telephone was still at last, "I'll meet you at the main information desk at United at eight-fifteen tomorrow morning, and we'll get you on that flight, OK?"

The TWA stewardess stopped by for the girl, and as they disappeared into the crowd Lenora put the small book she had been reading back into her purse. *Woodland Animals* was the title.

"I didn't think you were supposed to come to work till eight-thirty, Marlene," I said. "And don't you get to sleep late if you've stayed five hours over, the night before?"

She shrugged. "Eight-thirty, eight-fifteen. For fifteen minutes it's not going to kill me one way or the other."

"Eighty percent of the people in Kennedy Airport this minute," the information girl told me, "are lost. Some people get so nervous that they don't really think. And they don't know where they are going. And there are plenty of signs, but they don't read the signs . . ."

> BOARDING AREAS 1 THROUGH 7 INTERNATIONAL CONNECTIONS OBSERVATION DECK FLY THE FRIENDLY SKIES OF UNITED EXIT LOS ANGELES AIRPORT BUS STOP NEW YORK AIRWAYS HELICOPTER SERVICE FOR INFORMATION RESERVATION AND COURTESY BUS USE PHONES BEHIND DOOR DO NOT ENTER ARRIVING FLIGHTS DEPARTING FLIGHTS SPECIAL SERVICE FUTURE TICKETS NOTICE SNEAKERS ON ESCALATORS ARE DANGEROUS PERSONAS SIN BOLETAS NO MAS ALLA DE ESTE PUNTO METERED TAXI CABS LICENSED BY POLICE DEPARTMENT INTERAIRLINE COACH SERVICE TO ALL AIRLINES AT KENNEDY 25¢ LIMOUSINE AND CAR RENTALS INQUIRE AT COUNTER BETWEEN DOORWAYS A AND B FREE CONNECTION SERVICE FROM EAST SIDE AIRLINES TERMINAL

STAIRWAY TO UPPER LOBBY LOCATED BY TICKET COUNTER UNCLAIMED BAGGAGE WILL BE REMOVED TO THE BAGGAGE SERVICE OFFICE TO BOARDING AREA 1234567 STOP TAKE TICKET TICKETED PASSENGERS CHECK IN HERE FOR FLIGHTS 53, 311, 409 SE PROHIBE FUMAR DESPUES DE ESTE PUNTO HANGAR BUS ONLY RENTAL CAR PARKING USE EXTREME LEFT LANE NEW YORK BROOKLYN LONG ISLAND AND PARKING KEEP LEFT CLEARANCE 10′ 5″ SHELTER AREA PUSH GROUND TRANSPORTATION PULL DINING ROOM OPEN TILL 3 AEROFLOT MOCKBA TERMINAL CONNECTION BUS STOP EXPRESS TO LAGUARDIA SALA DE VISITANTES UNITED SKYPORT CINEMA TELEPHONE AHEAD FOR RESERVATIONS DISCOVER FLYING COCKTAIL LOUNGE OPEN FROM 1030 TILL MIDNIGHT US POSTAGE STAMPS COMPARE YOUR CLAIM CHECK SINCE MANY BAGS HAVE IDENTICAL APPEARANCE PLEASE COMPARE YOUR CLAIM CHECK WITH THE TAG ON YOUR BAG THANK YOU OFFICES TICKETS INFORMATION AND TICKETS TO MAKE FREE DIRECT CALLS 1. DEPRESS DESIRED NUMBERED BUTTON 2. LIFT RECEIVER, CONNECTION WILL BE COMPLETED IN CASE OF FIRE BREAK GLASS OPEN DOOR PULL HOOK TAXI CABS TIMES SQUARE $9 GRAND CENTRAL STATION $9 LAGUARDIA AIRPORT $4 POINTS OUTSIDE NEW YORK CITY FLAT RATE ONE TO 4-5 PERSONS BUS SERVICE TO GREENWICH RIVERSIDE STAMFORD DARIEN NORWALK WESTPORT BRIDGEPORT MILFORD NEW HAVEN MERIDEN AND HARTFORD FOR INFORMATION USE THIS DIRECT LINE NEW JERSEY LIMOUSINE SERVICE TRENTON WOODBRIDGE PRINCETON BERGEN COUNTY BRUNSWICK NEWARK AIRPORT WESTCHESTER LIMOUSINES TO NEW ROCHELLE WHITE PLAINS TARRYTOWN AND RYE ROCKLAND COUNTY TO NYACK AND SPRING VALLEY TRAVELERS AID PLEASE ENTER LOST AND FOUND FLIGHT INSURANCE JFK GROUND COMMUNICATIONS COCKTAIL LOUNGE OFFICES PLEASE STAND IN CENTER OF TREAD AND STEP OFF LAST STEP PLEASE HOLD HANDRAIL VISIT OUR HORIZON ROOM FOR COCKTAILS LUNCH AND DINNER WEATHER INFORMATION FLIGHT INFORMATION EXIT EXIT EXIT PARKING LOT NUMBER 3 ARRIVING PASSENGERS ON UPPER LEVEL CROSSWALK PRIVATE PROPERTY NO UNAUTHORIZED PARKING TOWAWAY ZONE WALKWAY DID YOU LOCK YOUR CAR? PUBLIC STENOGRAPHER NO

SMOKING BEYOND THIS POINT COIN CHANGER SHELTER AREA PUSH AUTOMATIC GATE PEDESTRIANS KEEP CLEAR LANES OPEN DO NOT ENTER CONCOURSE AIRPORT EXIT BANK CURRENCY EXCHANGE INFORMATION CASHIER ENTER STANDBY A B C PASSENGERS FROM FLIGHTS MARKED ARRIVE IN CLAIM AREA LOWER LEVEL NO STOPPING THIS IS NOT A PICKUP AREA MOTOR STAIR SNACK BAR EMERGENCY STOP TIMES SHOWN ARE SUBJECT TO CHANGE: INDICATED PM FLIGHT INFORMATION IS FURNISHED BY THE AIRLINES FOR FLIGHTS NOT LISTED SEE AIRLINE ON FIRST FLOOR HEAR PILOTS TOWER RADIO 10¢ ONE DIME OR TWO NICKELS SWITCH TO YOUR CHOICE AFTER CLEARANCE PASSENGERS EXIT TO LOBBY FIRST FLOOR INFORMATION DEUTSCH ESPANOL FRANCAIS ITALIANO WALKWAY TO AIR CANADA NATIONAL TRANSCARIBBEAN AUTHORIZED BUSSES ONLY 2 INTERNATIONAL ARRIVALS BUILDING 3 LOADING LAS VEGAS LISBON LONDON ROME PARIS CLEVELAND LOS ANGELES SAN FRANCISCO MADRID CHICAGO OAKLAND BOSTON ST LOUIS TEL AVIV ATHENS CINCINNATI OUT OF ORDER AUTOMATIC GATE TAKE TICKET TAX FREE GIFTS ALL AIRLINES MAIL POSTE TAX FREE LIQUOR 322 323 PARKING AT ANY TIME STOP YIELD TO DEPARTURES ARRIVALS NEXT LEFT 150TH ST. CARGO AREA NORTH PASSENGER TERMINAL TAXI HOLD AREA TAXIS ONLY FOR YOUR CONVENIENCE WE ARE EXPANDING THE INTERNATIONAL ARRIVAL AND WING BUILDINGS THE PORT OF NEW YORK AUTHORITY UNAUTHORIZED VEHICLES WILL BE TOWED AWAY AT OWNERS EXPENSE RESTRICTED TO TWA PASSENGER UNLOADING NO PARKING CURBSIDE CHECKIN BAGGAGE CHECKED HERE TELEPHONES TO PLANES PASSENGERS WITH TICKETS AN EXHIBIT OF ARTS AND CRAFTS BY NEW YORK BASES TWA CABIN ATTENDANTS ON THE BRIDGE LEVEL GATES 8-15 PLEASE PASSENGERS ONLY BEYOND THIS POINT LOCKERS BOOTBLACK NEWSPAPERS OF THE WORLD EMPLOYEES SHUTTLE SERVICE LOT NUMBER 7 PARKING FIELD REFLECTING POOL CONTROL TOWER DON'T WALK USE CROSSWALK TO PARKING LOTS AND QUICK PICKUP AREAS OUT ENTER BUS STOP NO STANDING EAST WING BUILDING DEPARTURES MERGING TRAFFIC Q-10 PUBLIC BUS WALKWAY PUSH BUTTON FOR WALK SIGNAL CAR LOADING SABENA LOFTLEIDIR CAUTION TRUCKS MEN AT WORK BUS TO NEW YORK

CITY PASSENGERS WITH TICKETS ONLY BEYOND THIS POINT.

There are plenty of signs, but they don't read the signs.

Kennedy Airport is an aquarium. It has been built at the bottom of an enormous ocean and we come to it in little air-filled vehicles and quickly enter air-filled chambers, completely self-sufficient undersea; each with its own coffee shops, restaurants, bookstores, resting places, viewing places out upon the sunken plains of a watery universe.

In from that universe come the fish of this ocean, sifting down from upper levels—turning, settling, iridescent hues shimmering in the liquids about them. Gold and silver, red and orange and green and black, salt-water tropicals grown a thousand times, hundred-ton angelfish, half-million-pound demoiselles angling in front of our view-ports, different sizes and shapes and colors, each family of fish clustering at its own feeding place.

Longer than locomotives, most of them, monstrous swept fins fifty feet, seventy feet high, they move ponderous and slow, infinitely patient, each to its own special grotto. They are gentle maneaters all, that can swallow a hundred or three hundred Jonahs more or less fearful of destiny, trusting the great fish to remain friend for just one more journey.

The fish themselves are unafraid. Giant leviathan noses loom right to our glass and we can look into the eyes and see purpose and motion there, we can watch the fish thinking, making ready another ocean-spanning continent-leaping voyage.

When the last Jonah is sealed inside, gills breathe, flukes move. The creatures scull ever so gingerly, turn, showing their colors and markings, and drift out to a place where they know there's room for the long arrowing thrust of their lift from the ocean floor.

We see them small in the liquid distance begin their push, bring their aquatic minds to focus on this drive, forget all else, force, blast their way into torrents of sea wind, surge free of the bottom in a cloud of rolling silt, curve bright-flashing toward the top of the sea, choose their turns, find their ways, settle up toward their far horizon and out of sight in the blue.

Coming, going, carefully releasing the world's Jonahs and carefully taking them aboard, the deep-sounding planet-traveling fish come to be known in time, by the people who watch. Some of the watchers are expert, having memorized Latin names, habit and habitat.

Others know only that these are mighty big fish, hope to tell you.

It used to be, years ago, before airplanes had radios and when the first control towers were built, that each tower had its "biscuit gun" with which a controller could shoot a colored beam of light to a pilot in his airplane, to advise him of what the tower man thought he should do. Flashing green: cleared to taxi. Steady red: stop. Steady green: cleared to land.

Today all that communicating is done by some crackerjack radio equipment, which all works very well. After an airline has spent three thousand dollars on a radio, natural-

ly, they expect that it will work very well.

Nevertheless, the first sight that caught my eye as I climbed the last flight of stairs to the glass aerie of the control tower at Kennedy Airport was the biscuit gun, suspended by a pulley wire from the ceiling. It hung there perfectly still, and there was dust on it.

Waist-high around this room, which is about twenty feet square, are radio and radar consoles, banks of switches to control the runway lights, communications to air traffic control rooms, telautographs for weather sequences, dials for wind speed and direction. (It has always seemed odd to me that a hundred-ton airliner still arranges itself in the sky so that it will land into the wind. One might expect that we'd become indifferent to a spirit so insubstantial as the wind, but not so.)

In this room stand five men, four of them young men and one old-timer, the watch supervisor, sitting back at his desk while the others stand, looking down on their kingdom Kennedy.

It is just before noon of a murky day and the mist has settled in a gray bowl over us. Just visible to the east is Jamaica Bay, same to the south, beyond runway 13 Right. To the north and west we can see to the edge of the airport and no farther.

The tower is the peak of a maypole, with airliners taxiing in a circle around it on the curved perimeter taxiways—clockwise on the south, counterclockwise on the north side of the tower, all converging on a path that leads to the end of 13 Right, the takeoff runway. Its sister, runway 13 Left, is for landings only, and for now there is practically nobody landing; 13 Left is a deserted wallflower of a runway, and looks lonely out there in the fog.

The airplanes blazing by on takeoff bellow up into exaggerated steep climbs, and I can't help but wince, watching them claw for altitude. That is maximum performance, the pilot is earning his keep on that kind of takeoff, and the planes disappear into the murk with their noses forced unnaturally high.

There is a twenty-minute delay for departure now, a twenty-minute wait in line for takeoff, but there is no tension in the tower. There is time for the younger ones to talk of who will be taking vacations when, time for yawning, time for the lighting of cigarettes in this air-conditioned cube.

Way down on the ground the fountains of the reflecting pool have been turned off. There are spaces in the parking lots. Along the ring of terminal buildings surrounding us I count a sparse forest of construction derricks at work: three in the new area north of BOAC, four at National, three at TWA, two at Pan American as they add extensions for their big new airplanes. In all there are fifteen cranes at work, lifting concrete in buckets and steel in beams.

The supervisor, the old-timer, opens a crinkly white bag and lays out three large ham-on-rye sandwiches on his desk. The ground controller, who talks to all the airplanes taxiing, calls across to him.

"Eastern wants to know the delay outbound. Got a new figure?"

"Well, there's six . . ." says the supervisor to himself, then, "Tell 'em half an hour."

The ground controller presses the button of his microphone. "Eastern 330, it'll be a half an hour approximate delay."

Each of the controllers wears earphones tuned to his own radio frequency, so I couldn't hear what Eastern 330 said to that. "Ah, roj," he probably said.

"That's a good sandwich," the supervisor says reflectively, for the quiet consideration of all. His words open a discussion on the construction of sandwiches, on lunches in general, on Chicken Delight, on franks and beans.

There are four radar screens in the tower.

And a copy of the *New York Post*.

And the door opens below and a man saunters up the stairs, unhurried, chewing a toothpick.

"There you are, Johnny," says the ground controller. "Thought I was going to go without lunch today."

The lunchward-bound takes a moment to tell his relief which airplanes are where, and hands him the microphone. The relief nods, opens a soft-drink can, chewing all the while on that toothpick.

Way off at the edge of the mist, there's a 707 touching down on 13 Left.

From here, the TWA terminal looks like the head and eyes of an enormous wasp, mandibles open, wings and body buried in the sand. It is watching the tower.

There are twenty airplanes waiting in line for takeoff.

"Here you go, Johnny-baby," says the departure controller, handing a strip of paper marked with numbers.

"Hm. Another Hugenot," Johnny-baby replies, looking at the numbers. "They're gatherin' at the gates."

"Say, Bob, we're going to run out of room here, with all these Hugenots . . . American 183, sir, you'll have to turn around here, that portion of the taxiway is all closed."

Down on the outer perimeter a 727 Trijet slows to a halt, then turns in cramped slow motion. A hundred yards ahead of him the taxiway is a rilled mass of bare earth, with graders combing it back and forth, back and forth.

"I wish they'd give us the airport back," Johnny says. "Let's call it forty minutes. Forty minutes delay . . ."

By the time I left the tower there was an hour's delay, and the line for takeoff stood forty planes long.

Two quite separate kingdoms, this land of Kennedy. One is the Kingdom of the Passenger, wherein the customer rules and all bend to his wish. The passenger reigns over the ground outside, the concourses, the shops and services, Customs, ticket counter, airline offices, and the aftermost nine-tenths of every airplane, where stewardesses ply him with refreshment and confidence.

The other tenth of that airplane is the Kingdom of the Pilot. And pilots are fascinating stereotyped people. They are almost exclusively men who like flying more than anything else in all the world, who work on the flight decks of jet transports not out of a wish to help passengers reach their many ports but because they like to fly and they're good at their job, most of them, and they wouldn't be much use in any other job anywhere. The exceptions to the generality, the ones who could do well at other work, don't make the best pilots. They can follow the numbers, all right, but when real flying skill is required (as it is at rare intervals nowadays and getting rarer), they are foreigners in the sky.

The best pilots are the ones who began flying when they were boys, who come to their gold-braid caps from turbulent histories of failure and distress in the ground-bound affairs of men. Not having the temperament or ability to bear the discipline and boredom of college, they failed or quit and took to flying full-time, enlisting in the Air Corps or making it the hard way—sweeping hangar floors, pumping gas as apprentice aviators, dusting crops, flying passenger rides, instructing, knocking about the country from one airport to the next, at last deciding to try the airlines since there's nothing to lose, trying, and glory be, getting hired!

All pilots live the same sky the world around, but airline pilots have more trappings and live more rigidly than do any other kind; than even military pilots. They must shine their shoes, wear neckties, be kind to all passengers, follow each comany rule and Federal Air Regulation, never lose their temper.

In return for this, they receive (a) more money for less work than any tradesman anywhere, and, most important, (b) the privilege of flying excellent airplanes, without having to apologize to anybody.

Today the major airlines require college training of their pilot applicants, and so lose the best stick-and-rudder airmen to the nonscheduled airlines (who need better pilots anyway, to cope with a wider range of problems), to agricultural and corporate flying concerns. Why the college requirement, is unclear, since all that a zoology-trained pilot has to fall back upon is Ichthyology 201, while the life-trained pilot, whose ranks are legion but diminishing, flies his airplane home on knowing born of interest and love instead of company requirement.

The path between the kingdoms at Kennedy is at best one-way . . . no one walks the pilot's kingdom who is not a pilot. And the path is very nearly no-way. The best of airmen is notoriously ill-at-ease on the ground, unless he is talking about flying, which he usually is and so makes do.

You can see it in the pilots coming off duty at Kennedy, all conservative uniforms and round-billed caps, whatever nation their airline. You'll see them awkward, self-conscious, most of them, looking straight ahead, in a hurry to get out of the passengers' kingdom and into somewhere more comfortable.

Each is painfully aware of his alien status in the concourses and decorated halls. To each there is nothing so indecipherable as the man who could choose to be passenger instead of pilot, the one who would choose any life but flight, who can stay away from the airplanes, not think about them even, and yet be happy. Passengers are a different race of humans, and pilots stay as far from them as courteously possible. Ask a pilot someday how many real friends he has who are not pilots themselves, and he will be hard-pressed to think of a single one.

The pilot is blissfully unaffected by anything that happens at the airport which does not directly bear on his flying—as far as he is concerned, the passengers' kingdom doesn't really exist, though occasionally he will look at the people with a benign sort of paternal affection. His world is very pure, without cynics or amateurs, and it is very simple. Its realities center on his airplane and fan out to include wind speed and direction, temperature, visibility, runway conditions, navigation aids, air traffic clearance, destination- and alternate-airport weather. That about locks it up. There are other elements: seniority, the six months' physical examination, flight checks in the aircraft, but those are ancillary to his kingdom, not the core of it. If ground traffic is bumper-locked in ten thousand automobiles, if there is a construction workers' strike, if organized crime is sordid and everywhere, stealing millions annually from the airport, he is completely untouched. The pilot's only reality is his airplane and the forces that affect it in flight. That is why airline travel is the safest transportation in the history of man.

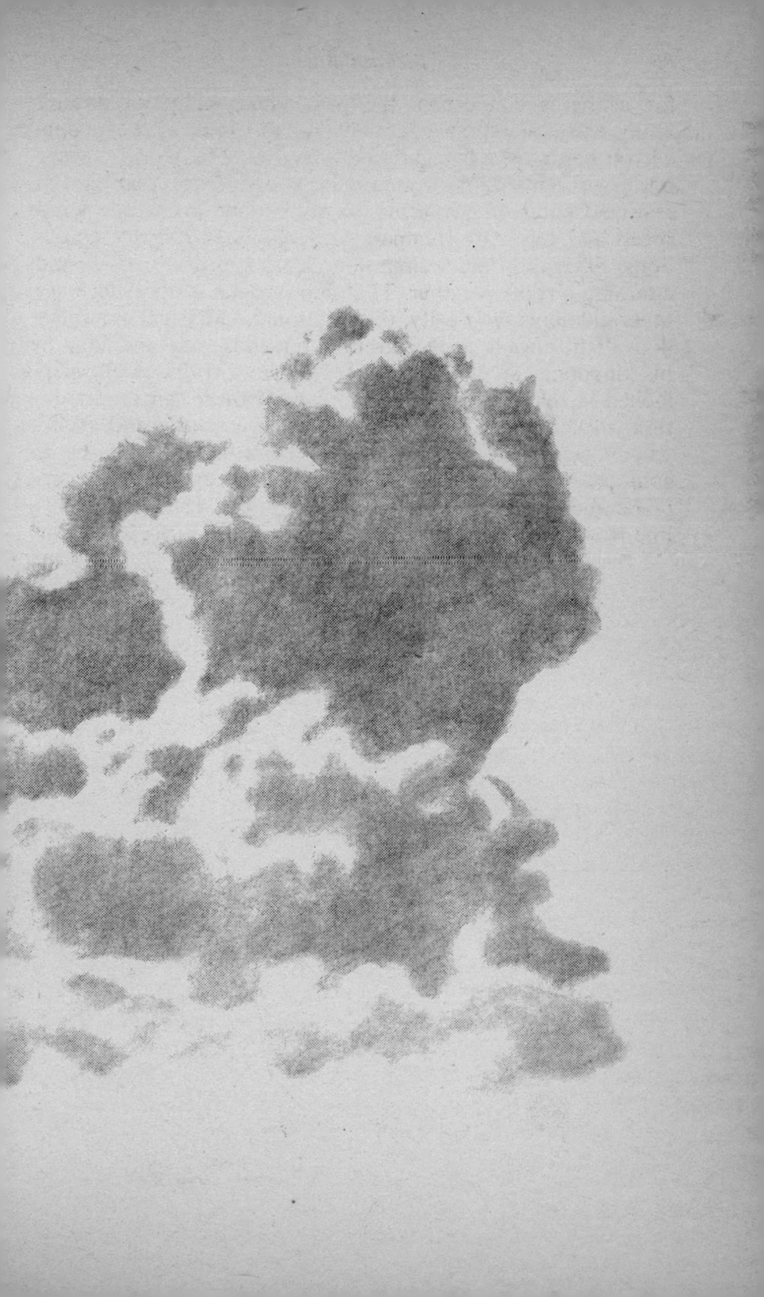

Perspective

I used to wonder, a few years ago, about railroad tracks. I'd stand between them, watch them go out into the world, and the two rails narrowed, they came together, they touched each other just five miles west, on the horizon. Monster locomotives would go hiss-thundering west through town, and since a locomotive is the kind of giant that needs its rails set just so, I knew there had to be a great pile of steaming wreckage just beyond the place where the tracks came together. I knew that the engineers had to be fiercely brave men, blurring past the Main Street crossing with a grin and a wave, facing certain death on the horizon.

Eventually, I found that the railroad tracks didn't really meet beyond our town, but I didn't get over my awe around railroad men till the day I met my first airplane. Since then, I've followed track all over the country and haven't yet seen a set of rails come together. Ever. Anywhere.

I used to wonder, a few years ago, about fog and rain: why was it, some days, that the whole earth was gray and wet, the whole world a miserable, flat, sad place to live? I wondered how bleakness happened to the whole planet at once, and how it was that the sun, so bright yesterday, had turned to ash. Books tried to explain, but it wasn't till I began to know an airplane that I found that clouds don't cover the whole world at all—that even from where I stood in the worst of the rain, soaking wet on the runway, all I had to do to find the sun again was to fly above the clouds.

It wasn't easy to do that. There were certain definite rules to follow, if I really wished to gain the freedom of clear air. If I chose to ignore those rules, if I chose to thrash around wildly, to insist that I could tell up from down all on my own, following the impulse of the body instead of the logic of understanding, I would invariably fall down. In order to find that sun, even today, I have to ignore what seems right to my eyes and hands, and rely totally on the instruments given me, no matter how strangely they seem to speak, how senseless they appear to be. Trusting those instruments is the only possible way anyone can ever break out into the sunlight. The thicker and darker the cloud, I found, the longer and more carefully I had to trust the pointers and my skill in knowing what they say. I proved it over and again: if only I kept climbing, I could reach the top of any storm, and lift into the sun at last.

I learned, when I began flying, that boundaries between countries, with all their little roads and gates and checkpoints and *Prohibited* signs, are quite difficult to see from the air. In fact, from altitude I couldn't even tell when I had flown across the border of one country into another, or what language was in fashion on the ground.

An airplane will bank to the right with right aileron, I found, no matter if it's American or Soviet, British or Chinese or French or Czech or German, no matter who's flying it, no matter what insignia is painted on the wing.

I've seen this and more, flying, and it all falls under one label. Perspective. It is perspective, it is getting above the railroad track, that shows we needn't fear for the safety of locomotives. It is perspective that shows us beyond the illusion of a sun's death, that suggests if we lift ourselves high enough, we'll realize that the sun has never left us at all. It is perspective that shows the barriers between men to be imaginary things, made real only by our own believing that barriers exist, by our own bowing and cringing and constant fear of their power to limit us.

It is perspective that stamps itself upon every person up for his first flight in an airplane: "Hey, the traffic down there . . . the cars look like *toys!*"

As he learns to fly, the pilot discovers that the cars down there *are* toys, after all. That the higher one climbs, the farther he sees; the less important are the affairs and crises of those who cling to the ground.

From time to time, then, as we walk our way on this little round planet, it's good to know that a lot of that way can be flown. We might even find, at the end of our journey, that the perspective we've found in flight means something more to us than all the dust-mote miles we've ever gone.

The pleasure of their company

"You'll want to press that little brass plunger there . . . flood the carburetor before she'll start."

It was a month into summer and a minute into sunrise. We stood at the edge of a sixteen-acre meadow, a mile north of Felixstowe on the Ipswich road. David Garnett's Gipsy Moth was fresh-dragged from her shed, wings unfolded and locked in place, tailskid hidden in grass. Across the field the first birds were coming awake, larks or something. There was no wind.

I pushed the plunger and the frail metal squeak of it was the only man-made sound of morning, until the petrol fell from the engine and hit the dark grass.

"You can take the rear cockpit, if you wish. I'm up for the ride," he said. "Careful of the compass, getting in. I've smashed the thing twice now, myself. If I wasn't right at home with it set down there on the floor, I'd throw it out and get a better. Switches off."

He stood by the propeller in his tweedy-cloth flying clothes, in no particular hurry, enjoying the morning.

"You do *have* switches in this machine, David?"

I felt like a dumb Colonial. Supposed to be an airplane pilot and I can't even find the magneto switch.

"Oh, yes. Sorry I didn't say. Outside the cockpit, next the windscreen. Up is on."

"Ah, so." I checked that they were down. "They say off."

He pulled the propeller through a couple of times, calm and easy, with the detachment of one who has done this a thousand times over and still enjoys it all. He had learned to fly rather late on in life, and it had taken him twenty-eight hours of dual instruction before he finally soloed the Moth. He neither brags nor apologizes over it. One of the best things about David Garnett is that he is honest with himself and the world, and therefore is a happy man.

"Switches on," he called.

I clicked them up. "OK. You're hot."

"Pardon?"

"Switches on."

He pulled the propeller quickly down with one hand and a practiced turn of wrist, and the engine caught at once. After a brief little roar it settled to perk quietly at 400 rpm, with the sound of a small inboard Chris-Craft at idle on a blue-morning lake.

Garnett climbed rather awkwardly into the front cockpit, fastened his leather helmet down over his head, and adjusted his Meyrowitz goggles—of which he is quite proud, for they are first-rate goggles. When he isn't flying his helmet and goggles hang on a hook just over his fireplace at Hilton.

I let the Gipsy engine warm for a few minutes, than touched the throttle forward and we scraped and teetered to face the longest way across the field. The Moth had no brakes, so I checked the magnetos quickly on takeoff, and, full power, the machine leaped up into the air.

It was a little like that moment in a spectacular motion picture when for visual effect they run the film in black and white, and then flick it over into color. As we came off the grass, the sun burst and sprayed yellow light all over England, which strangely made the trees and meadows go full, deep British green, and the lanes gold and warm.

I played about with the airplane a bit, a lazy eight and a steep turn, but most of all just little turns and a climb up to one thousand feet and a rush back down to sea level below the cliffs by the ocean, dodging gulls.

The haze came up an hour later, and clouds capped it down to earth, so we pulled up into the gray, keeping the airspeed between sixty and seventy and the sun overhead, till we broke out on top at three thousand feet, ". . . above a plain of vapour," as David would say. The sun shone brilliantly, black shadows of struts and wires striped the wings. We were alone with the cloud and with our thoughts that morning. Only an occasional triangle of green slid below to remind us that the earth still existed, somewhere.

At last I shut down the engine and duplicated a flight that he had told me about: ". . . yes, there were the hangars and the aerodrome . . . (and there they were, and two miles beyond, our meadow) . . . I did a big sideslip, but even so I overshot and went round again . . . (so did I—we were still two hundred feet up when we came across the fence) . . . This time my approach was perfect and my landing curiously soft and dreamlike. I was on the earth, but the earth was unreal, a limbo of haze and softened sunlight. Reality was far above me . . ."

I've done a lot of flying with this soft-spoken fellow, and in this day of few real friends, when a man is fortunate to go past three counting them, David Garnett is a real friend. We like the same things: the sky, the wind, the sun; and when you fly with somebody who puts his value on the same things that you do, you can say that he is a friend. Anyone else in that Moth, bored by the sky, would no more have been a friend than that businessman twelve rows down the aisle of a 707, though we share our flying a thousand times.

In a way, I know Garnett even better than his own wife knows him, for she can never quite understand why he wants to throw hours away in that noisy, windswept contraption that sprays oil all over one's face. I do understand why.

But probably the most curious thing about knowing David Garnett is that though we've done a lot of flying together and though I know him very well, I have no idea what the man looks like, or even if he is still alive. For David Garnett is not only an airplane pilot, he is a writer, and to one way of thinking, the talks we've had and the places we've flown have all been between the battered covers of his book, *A Rabbit in the Air,* published in London in 1932.

The way to know any writer, of course, is not to meet him in person but to read what he writes. Only in print is he most clear, most true, most honest. No matter what he might say in polite society, catering to convention, it is in his writing that we find the real man.

David Garnett, for instance, writes that after flying those twenty-eight hours of dual instruction, after flying those thirty-six lessons, all he did after his first flight alone in the Moth was to step out of the cockpit and smile and sign up for some more flying time. And that is all we would have seen, had we stood and watched him that Wednesday afternoon in the end of July, 1931, at Marshall's Aerodrome.

But was he really so unmoved by his first solo? We have to leave the aerodrome to find out.

"Half way home, I asked myself alone in the supercilious voice which has so often been used to me, 'Have you gone solo yet?'

" 'Yes.'

" 'Have you gone solo?'

" 'Yes!'

" 'Have you gone solo?'

" *'YES!'* "

Does that sound familiar? Remember when you were learning to fly, driving home after each lesson, that condescending pity you felt for all the other drivers, bound tightly as they were to their little cars and their little highways? "How many of *you* have been flying just now? How many of *you* have just looked away out across the horizon, have ten minutes ago won a battle with a fierce crosswind across a narrow runway? None of you, you say? You poor people . . . I HAVE," and pulling back the steering wheel of your automobile, you could almost feel her going light on her wheels.

If you remember that time, you have a friend in David Garnett, and to meet him costs a dollar or so in a secondhand bookstore.

Thousands of volumes have been written about aviation, but we do not automatically have thousands of true and special friends in their authors. That rare writer who comes alive on a page does it by giving of himself, by writing of meanings, and not just of fact or of things that have happened to him. The writers of flight who have done this are usually found together in a special section on private bookshelves.

There are rafts of flying books left from World War II, but nearly every one of them is absorbed in fact and exciting adventure, and the author shies away from the meaning of the fact, and of what the adventure stands for. Perhaps he is afraid to be thought egotistic, perhaps he has forgotten that each one of us, in the moment that one reaches toward a worthwhile goal, becomes a symbol of all mankind striving. In that moment, the word "I" doesn't mean a personal, egocentric David Garnett, it means all of us who have loved and wished and struggled to learn, and who have soloed our Moth at last.

There is something about a blend of fact and meaning and pure honesty that gives a book presence, that puts us in that cockpit, for better or worse, heading out to meet our destiny. And when you walk the same path toward destiny with a man, that man is likely to become your friend.

Out of World War II, for instance, we meet a pilot named Bert Stiles, in a book he called *Serenade to the Big Bird.* The Big Bird is a Boeing B-17 Flying Fortress, running combat missions out of England in France and Germany.

Flying with Bert Stiles turns us weary to death of war, and of eight hours a day in the right seat, sitting and wrestling with the airplane or sitting and doing nothing while the aircraft commander wrestles with it. The oxygen goes stale in our mask, the flak comes up all black and yellow and silent, the black-cross Messerschmitts and Focke-Wulfs come rolling through us in head-on attacks, yellow fire sparkling from their nose cannon and thuds and splinters through the plane and bombs away and the whole complete entire High Squadron is shot out of the air and a hard thud and orange flame from the right wing and pull the fire handle and feather Four and the Channel at last the beautiful Channel and straight in to land home on the ground and chow without taste and sack without sleep and right away Lieutenant Porada snapping on the light to say Come on, breakfast at two-thirty briefing at three-thirty and start engines and takeoff and sitting there in that right seat while the oxygen goes stale in our mask, the flak comes up all black and yellow and silent, the black-cross Messerschmitts and Focke-Wulfs come rolling through us in head-on attacks, yellow fire sparkling from their nose cannon . . .

Flying with Stiles, there is no glory, and a bomb run is not even flying. It is a dirty terrible job that's got to be done.

"It will be a long time before I have made up my mind about this war. I am an American. I was lucky enough to be born below the mountains of Colorado. But someday I would like to be able to say I live in the world and let it go at that.

"If I live through this, I will have to get on the ball and learn something about economics and people and things . . . In the end it is only people that count, all the people in the whole world. Any land is beautiful to someone, any land is worth fighting for to someone. So it isn't the land. It is the people. That is what the war is about, I think. Beyond that I can't go very far."

After his combat tour in bombers, Stiles volunteered to fly combat in P-51s. On November 21, 1944, he was shot down on an escort mission to Hanover. He was killed at age twenty-three.

But Bert Stiles did not die before he had a chance to arrange some patterns of ink on two hundred pieces of paper, and in that arranging he has become a voice inside our head and sight inside our eyes to see and to wonder and to talk honestly about his own life and therefore about ours.

The only important part of Bert Stiles was set to paper near an Eighth Air Force runway those thirty years ago, and that same paper is here for us to touch and know and see within, this minute. That important part is what makes any man what he is and what he means.

To talk in person with Antoine de Saint-Exupéry, for instance, we would have had to peer through a constant cloud of cigarette smoke about his head. We would have had to listen to him worry over imaginary diseases. We would have had to stand at the airport and wonder . . . would he remember to lower the landing gear today?

But as soon as Saint-Exupéry ran out of excuses not to write (and these were many), as soon as he found his inkwell amid the clutter of his room and when at last his pen touched paper, he set free some of the most moving and beautiful ideas about flight and man that have ever been written. Few are the pilots, reading his thought, who cannot nod and say, "That's true," who cannot call him friend.

" 'Careful of that brook (said Guillamet), it breaks up the whole field. Mark it on your map.' Ah, I was to remember that serpent in the grass near Mortril! Stretching its length along the grasses in the paradise of that emergency landing field, it lay in wait for me a thousand miles from where I sat. Given the chance, it would transform me into a flaming candelabra. And those thirty valorous sheep ready to charge me on the slope of a hill.

" 'You think the meadow empty, and suddenly bang! there are thirty sheep in your wheels.' An astounded smile was all I could summon in the face of so cruel a threat . . ."

In the very best among the writers of flight, we might expect to find some very lofty and difficult thought set to paper. But not so. In fact, the higher the quality of the writer, and the better a friend he becomes to us, the more simple and clear is the message that he brings. And strangely, it is a message that we do not learn as much as remember, something we find that we have always known.

In *The Little Prince,* Saint-Exupéry lays out the idea of this special kind of friendship that airplane pilots can have with other pilots who have written of flight.

" 'Here is my secret,' said the fox to the little prince, 'a very simple secret: It is only with the heart that one can see rightly; what is essential is invisible to the eye.'

" 'What is essential is invisible to the eye,' the little prince repeated, so that he would be sure to remember."

Saint-Ex writes of you and me, who are drawn to flight in the same way that he was drawn to it, and we look for the same friends within it. Without seeing that invisible, without recognizing that we have more in common with Saint-Exupéry and David Garnett and Bert Stiles and Richard Hillary and Ernest Gann than we have with our next-door neighbor, we have left them all untamed, and they are no more friends than a hundred thousand unknown faces are friends. But as we get to know that real man who is set down on paper, that man to whom the living mortal devoted his lifetime, each of these becomes, for us, unique in all the world. What is essential about them, and about us, is not seen with eyes.

We are friend to a man not because he has brown hair or blue eyes or a scar on his chin from an old airplane crash, but because he dreams the same dreams, because he loves the same good and hates the same evil. Because he likes to listen to the sound of an engine ticking over on a warm, quiet morning.

Facts alone are meaningless.

FACT: The man who wore the uniform of commandant in the French Air Force, who carried a flight log written with seven thousand hours and the name *Saint-Exupéry,* did not return from a reconnaissance flight over his homeland.

FACT: Luftwaffe Intelligence officer Hermann Korth, on the evening of July 31, 1944, the evening when Saint-Exupéry's was the only aircraft missing, copies a message —"Report by telephone . . . destruction of a reconnaissance plane which fell in flames into the sea."

FACT: Hermann Korth's library in Aix-la-Chapelle, with its honored shelf for the books of Saint-Exupéry, was destroyed by Allied bombs.

FACT: None of this destroyed Saint-Exupéry. Not bullets in his engine or flames in his cockpit or bombs tearing his books to shreds, for the real Saint-Ex, the real David Garnett, the real Bert Stiles are not flesh and they are not paper. They are a special way of thinking, much like our own way of thinking, perhaps, but still, like our prince's fox, unique in all the world.

And *meaning?*

These men, the only part of them that is real and lasting, are alive today. If we seek them out, we can watch with them and laugh with them and learn with them. Their logbooks melt into ours, and our flying and our living grows richer for knowing them.

The only way that these men can die is for them to be utterly forgotten. We must do for our friends what they have done for us—we must help them to live. On a chance that you may not have met one or two of them, will you allow me the honor of introductions?

MR. HARALD PENROSE, *No Echo in the Sky* (Arno Press, Inc.)

MR. RICHARD HILLARY, *The Last Enemy* (also published with the title *Falling through Space*)

FLT. LT. JAMES LIEWELLEN RHYS, *England Is My Village* (Books for Libraries, Inc.)

MRS. MOLLY BERNHEIM, *A Sky of My Own* (Macmillan Publishing Co., Inc.)

MR. ROALD DAHL, *Over to You*

MISS DOT LEMON, *One-One*

SIR FRANCIS CHICHESTER, *Alone over the Tasman Sea*

MR. GILL ROBB WILSON, *The Airman's World*

MR. CHARLES A. LINDBERGH, *The Spirit of St. Louis* (Charles Scribner's Sons)

MRS. ANNE MORROW LINDBERGH, *North to the Orient* (Harcourt Brace Jovanovich, Inc.)

MR. NEVIL SHUTE, *Round the Bend, The Rainbow and the Rose, Pastoral* (Ballantine Books, Inc.)

MR. GUY MURCHIE, *Song of the Sky* (Houghton Mifflin Company)

MR. ERNEST K. GANN, *Blaze of Noon* (Ballantine Books, Inc.), *Fate Is the Hunter* (Simon & Schuster, Inc., Ballantine Books, Inc.)

MR. ANTOINE DE SAINT EXUPERY, *Wind, Sand and Stars* (Harcourt Brace Jovanovich, Inc.), *The Little Prince* (Harcourt Brace Jovanovich, Inc.)

If the book is in print, the publisher is listed. Otherwise look in libraries and secondhand bookstores.

A light in the toolbox

That which a man believes, the philosophers say, is that which becomes his reality. So it was for years as I said over and again "I'm no mechanic," I was no mechanic. As I said "I don't even know which end of the screwdriver to hit the nail with," I closed a whole world of light from myself. There had to be somebody else to work on my airplanes, or I couldn't fly.

Then I came to own a crazy old biplane, with an old-fashioned round engine on its nose, and it didn't take long to discover that this machine was not about to tolerate a pilot who didn't know something of the personality in a 175 horsepower Wright Whirlwind, something about the repair of wooden ribs and doped fabric.

That was how the rarest event in life came to me . . . I changed the way I thought. I learned the mechanics of airplanes.

What everybody else had known for so long was brand-new adventure to me. An engine, for instance, torn apart and scattered across a workbench, is just a collection of odd-shaped pieces, it is cold dead iron. Yet those same pieces, assembled and bolted into a cold dead airframe, become a new being, a finished sculpture, an art-form worthy of any gallery on earth. And like no other sculpture in the history of art, the dead engine and the dead airframe come to life at the touch of a pilot's hand, and join their life with his own. Standing separately, the iron and the wood and the cloth and the man are chained to the ground. Together, they can lift on up into the sky, exploring places where none of them has ever been before. This was surprising for me to learn, because I had always thought that mechanics was broken metal and muttered curses.

It was all there in the hangar to see, the moment I opened my eyes, like an exhibit in a museum when the light is turned on. I saw on the bench the elegance of a half-inch socket set; the smooth, simple grace of an end-wrench, wiped clean of oil. Like a new art student who in one day first sees the work of Vincent Van Gogh and Auguste Rodin and Alexander Calder, so I suddenly noticed the work of Snap-On and Craftsman and the Crescent Tool Company, gleaming silent and waiting in battered toolbox trays.

Art of tools led to art of engines, and in time I came to understand the Whirlwind, to think of it as a living friend with whims and fancies, instead of a mystic sinister unknown. What a discovery that was, to find what was going on inside that gray steel case, behind the spinning flash of the propeller blade and the flickering bursts of engine roar. No longer was it dark inside those cylinders, around that crankshaft; there was light—I knew! There was intake and compression and power and exhaust. There were pressure oil bearings to hold whirring high-speed shafts; carefree in-

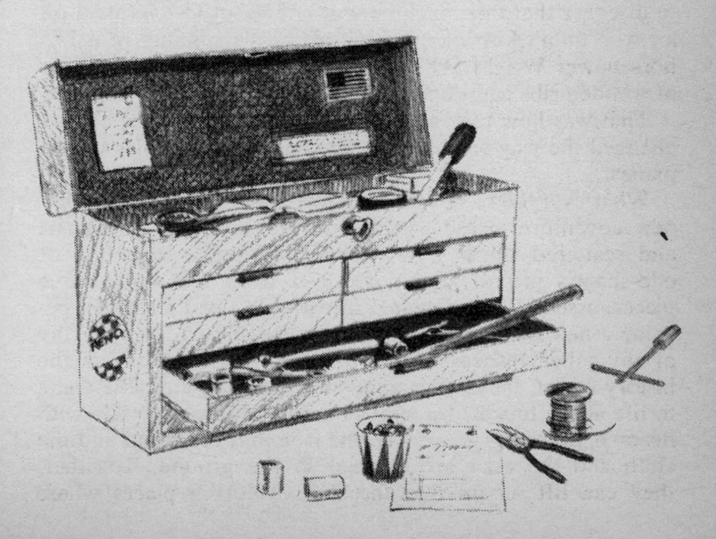

take valves and tortured exhaust valves darting down and back on microsecond schedules, pouring and drinking fresh fire. There was the frail impeller of the supercharger, humming seven times round for every turn of the propeller. Rods and pistons, cam-rings and rocker-arms, all began to make sense, clicking to the same simple, straight logic of the tools that had bolted them in place.

I went from engines to airframes in my studies, and learned about weld clusters and bulkheads, stringers and rib-stitching, pulleys and fair-leads, wash-in, offsets, rigging. I had been flying for years, and yet this was the first day I ever saw an airplane, studied it and noticed it. All these little parts, fitting together to make a complete aircraft—it was great! I raged in the need to own a field full of airplanes, because they were so pretty. I needed them so that I could walk around and look at them from a hundred different angles, in a thousand lights of dawn and dark.

I began buying my own tools, began keeping them on my desk, just to look at and touch, from time to time. The discovery of the mechanics of flight is no small discovery. I spent hours in the hangar absorbing Michelangelo airplanes, in shops studying Renoir toolboxes.

The highest art form of all is a human being in control of himself and his airplane in flight, urging the spirit of a machine to match his own. Yet I learned, courtesy of a crazy old biplane, that to see beauty and to find art I don't have to fly every moment of my life. I have only to feel the satin metal of a nine-sixteenths-inch end-wrench, to walk through a quiet hanger, simply to open my eyes to the magnificent nuts and bolts that have been so close to me for so long.

What strange, brilliant creations are tools and engines and airplanes and men, when the light is turned on!

Anywhere is okay

It was just as if somebody had thrown a hundred-pound firecracker, had lighted the fuse two hours after midnight, thrown the thing high in the dark over our airplanes, over us asleep there in the hay, and run like crazy.

A ball of dynamite fire shattered us alive, bullets of hard rain burst like hail across our bedrolls, black winds tore us like animals gone wild. Our three planes leaped frantic against their ropes, strained up hard against them, tugged and kicked and clawed at them mad to go tumbling in the night with that maniac wind.

"Get the strut, Joe!"

"What?" His voice was blown away in wind, drowned in rain and thunder. In the lightning flash he was frozen the color of ten million volts, as were trees, leaves flying off, and the horizontal raindrops.

"THE STRUT! GRAB THE STRUT AND HANG ON!"

He threw his weight on the wing in the instant the storm snapped branches from the trees—between us we held the Cub from taking both of us under its wings and cartwheeling across the valley.

Joe Giovenco, a hippy teenager from Hicksville, Long Island, from the shadow of New York City, whose total understanding of thunderstorms had been that they made faint rumblings beyond the city in summertime, clung to that strut with python strength, personally battling wind and lightning and rain, his hair blowing in fierce dark tangles about his face and shoulders.

"MAN!" he shouted a second before the next dynamite went off, "I'M REALLY LEARNING A LOT ABOUT METEOROLOGY!"

In half an hour the storm rolled by and left us a warm dark calm. Though we saw the sky flickering and crunching in the hills to the east, and though we looked warily west for other lightnings, the calm stayed and we snaked at last back into our ragged wet bedrolls. Moistly though we slept, there was not one of the six of us out there in the night who didn't count the Invitational Cross-Country Adventure with the best wishes of his life. Yet it was nothing tried against great odds. All that brought us to it, or it to us, was that we shared a certain curiosity about the other people who live on our planet and in our time.

Maybe the headlines started the Adventure, or the magazine articles or the radio news. With their ceaseless talk of alien youth and generation gaps grown into uncrossable deep chasms and the only hope the kids have for the country is to tear it down and not rebuild at all . . . maybe that's where it started. But considering all this, I found that I didn't know any such kids, didn't know anybody unwilling to talk to those of us who were kids ourselves, yesterday. I knew there was something to say to one who says "Peace" instead of "Hello," but I didn't know quite what that might be.

What would happen, I thought, if a man with a little cloth-wing airplane came down to land on a road to offer a knapsacked hitchhiker a ride? Or better, what would happen if a couple of pilots made room in their planes for a couple of city kids for a flight of a hundred miles or so, or a thousand miles or so; a flight of a week or two across the hills and farms and plains of America? Kids who have never seen the country before, outside their high school fence or expressway overpass?

Who would change, the kids or the pilots? Or would both, and what kind of change might that be? Where would their lives touch, and where would they be so far apart that there could be no calling across the gulf?

The only way to find out what can happen to an idea is to test it, and that is how the Invitational Cross-Country Adventure came to be.

The first day of August, 1971, was a misty dim day—afternoon, in fact, by the time I landed at Sussex Airport, New Jersey, to meet the others.

Louis Levner owned a 1946 Taylorcraft and liked the idea of the flight sight unseen. For a target we chose the EAA Fly-in at Oshkosh, Wisconsin, reason enough to fly even if everybody else canceled out at the last minute.

Glenn and Michelle Norman of Toronto, Canada, heard about the flight, and though they weren't quite hippy kids they were strangers to the United States, eager to see the country in their 1940 Luscombe. And waiting there at the field when I landed were two young men who had labeled themselves Hippy for all the world to see. Hair down to their shoulders, headbands made of rags, dressed in faded dungarees, knapsacks and bedrolls at their feet.

Christopher Kask, thoughtful, nonviolent, almost non-speaking among strangers, had won a Regents scholarship out of high school, a distinction reserved for the top two percent of the student body. He wasn't sure, however, that college is America's best friend, and to get a degree for the sake of finding a better job did not sound to him like real education.

Joseph Giovenco, taller, more open with others, noticed everything with a careful photographer's eye. He knew there was a future in video tape as an art form, and video tape he'd be learning, come fall.

None of us knew just what would happen, but the flying sounded like fun. We met at Sussex and we cast anxious looks at the sky, at the mists and clouds there, not saying much because we were not yet sure how to talk with each other. We nodded at last, packed our bedrolls aboard, started engines, and rolled fast along the runway and into the sky. Over the noise of the engines, there was no way to tell what the kids thought, airborne.

My own thought was that we weren't going to get very far in the first flight. Clouds swirled up deep gray broths on the ridges west, with chunks of fog steaming in the branches of the trees. Blocked to the west, we flew south for ten miles, for fifteen, and finally, with the soup boiling and thickening all about us, came down on a little grass strip near Andover, New Jersey.

In the silence of that place, the rain began ever so gently to fall.

"Not what you might call an auspicious start," somebody said.

But the kids were undampened. "All the *land* in New Jersey!" said Joe. "I thought it was *populated!*"

I hummed the tune to *Mosquitoes, Stay Away from My Door* as I unrolled my blanket in the grass, glad that we weren't all gloomed by the terrible weather, hoping that tomorrow would dawn bright and see us on our way over our horizons.

It rained all night long. Rain with the sound of gravel pouring on drum-fabric wings, thudding into grass dryly at first, then with splashings as grass became marsh. By midnight we had given up hope of any star or of any sleep in the marsh; by one a.m. we were huddled and folded into the airplanes, trying at least to doze. At three a.m., after hours without a word, Joe said. "I have never been in rain this hard in my life."

Dawn was late, because of the fog . . . we had fog and clouds and rain for four days straight. In four days of taking to the air with every small break in the sky, in four days of dodging rainstorms and detouring them and hopping from one little airport to another we had flown a grand total of sixty-two miles toward Oshkosh, one thousand miles away. We slept in a hangar at Stroudsburg, Pennsylvania; in an airport office at Pocono Mountain; in a flying-school clubhouse in Lehighton.

We decided to keep a journal of the flight. Out of this, and out of our talks under the rains and amid the fogs, we began to know each other, ever so slightly.

Joe was convinced right off, for instance, that airplanes had personalities, that they had characters like people, and he didn't mind saying that the blue-and-white one over there in the corner of the hangar made him nervous. "I don't know why. It's the way it sits there looking at me. I don't like it."

The pilots jumped on that and told stories of airplanes that lived in different ways and did things that couldn't be done—took off in impossibly short distances when they had to, to save somebody's life, or glided impossibly long ones with engines stopped over jagged lands. Then there was talk about the way wings work, and flight controls and engines and propellers, and then about crowded schools and drugs on campus, then of how it is that sooner or later what a person holds fast in his thought becomes true in his life. Outside, the black rain; inside, the echo and murmur of voices.

In the journal we wrote whatever we didn't feel like saying aloud.

"This is really something!" Chris Kask wrote on the fourth day. "Every day is a string of surprises—some really *unbelievable* things have been happening. A guy lends us his Mustang, a guy lends us his *Cadillac*, everybody's letting us sleep in the airports and really going out of their way to be nice. It doesn't matter where we are or if we ever get to Oshkosh. Anywhere is okay."

The kindness of people was something the kids couldn't believe.

"I used to walk with Chris in a store or follow him down a street," Joe said, "and watch people watching him. His hair was as long as it is now—longer. They'd pass him and they'd look, sometimes they'd even stop and make some face or some remark. Condemn him. You could see the distaste in their eyes, and they didn't even know who he was!"

After that I took to watching people watching our hippies. Always there was a shock there, seeing them for the first time, the same startlement I had felt when I first saw them. But if either of them had a chance to talk, though, a chance to show that they were gentle people who did not plan to whip out bombs and blow everybody to pieces, that flicker of hostility vanished in something less than half a minute.

Once we were trapped by weather over the ridges of western Pennsylvania. We fell back from it, then circled and landed our planes in a long field of mown hay by the town of New Mahoning.

Scarcely had we stepped out onto the ground when the farmer arrived, his pickup truck rolling soft and crunchy on the wet stubble.

"Having some trouble, are you?" He said that first, and then he frowned when he saw the kids.

"No, sir," I said. "A little. The clouds were getting a bit low and we thought it might be better to land than to maybe fly into a hill up there. Hope you don't mind . . ."

He nodded. "It's OK. Everybody's all right, are they?"

"Thanks to your field. We're fine."

In minutes three other trucks and a car nosed down the dirt lane and onto the field; there was curious lively talk everywhere.

". . . saw them flying low over Nilsson's place there, and I figure he was in trouble. Then the two others come around and they went down and it got quiet and I didn't know what was going on!"

All the farm people with haircuts, all of them smooth-shaven, they flickered their eyes over the long hair and the headbands and they weren't sure what they had, here.

Then they heard what Joe Giovenco was saying to Nilsson.

"Is this a farm? A real farm? I've never seen a real . . . I'm from the city . . . that isn't corn is it, growing out of the ground?"

Frowns vanished in smiles like slow candles lighting.

"Sure that's corn, son, and that's the way it grows, right there. Sometimes you worry. This rain, now. Too much rain, and then a big wind right after, and the whole crop gets knocked flat and you've got troubles, sure enough . . ."

Somehow, that was a good scene to watch.

You could see the thoughts in their eyes. The hippies a fellow sets his jaw against are the sullen ones that don't care about the rain or the sun or the land or the corn . . the ones that don't do anything but cut the country down. But these kids, now, they're not that kind—a man can tell that right away.

When the ridges cleared, we offered rides in the airplanes, but no one was quite ready to go up. We started engines, then, and bounced up from the hay into the sky, rocked our wings farewell and went our way.

"Amazing!" Chris wrote in the journal that night. "We landed in a field and talked to farmers with Swedish and Irish accents—I didn't know this existed in Pennsylvania. Everybody is so nice. Friendly. It's really opened my eyes. A lot of my natural defenses are broken down. Just don't worry and trust things to work out. All my little plans for the future have really been shaken. I'm just not sure of anything anymore and that's good because it teaches you to go with the flow of things."

From that day we wafted west in pure blue air over the pure green land and farms like sunlight growing.

After all our explaining on the ground, Chris and Joe were ready to take the controls themselves. Their first hours of dual instructions were given in formation flying.

"Small corrections, Joe, SMALL CORRECTIONS! You want to hold the other plane just about . . . there. OK? You've got it, you're flying. Small corrections, now. Add a little power, close it in a little. SMALL CORRECTIONS!"

Before many hours were gone, they could actually hold

the airplanes in formation with each other. It was hard work for them, they made it much harder than it had to be, but still they soaked it in and waited like vultures after takeoff to pounce on the controls and practice some more.

Next they began making takeoffs themselves . . . squirrely weaving panic-stricken disasters at first, leaping in the last instant over runway lights and snow markers along the sides of the strip. When they got smoother, we practiced stalls and a spin or two coming down from formation, and at last they began making landings, learning, absorbing like sponges dropped in the sea.

Every day, too, we learned of their life and their language. We practiced talking Hippy, my notebook becoming a dictionary of that tongue. Joe insisted that I had to slur my words much more carefully—we practiced saying "Hey, man, what's happening?" over and over again, but it was harder than formation flying . . . I never did get it right.

" 'You know,' " Joe said, "means 'Um' or 'Duh.' 'Right on' means 'I agree emphatically,' said only to obvious statements and usually said by dummos."

"What is it," I asked, "when you 'make the scene'?"

"I don't know. I've never made it."

Though my dictionary had much about the language of drugs (marijuana is also Mary Jane, grass, pot, stuff, smoke, and Cannabis sativa; a 'nick' is a five-dollar bag of grass, 'spaced out' is how one feels while smoking it), neither of the kids had brought any along on the Cross-Country Adventure. This puzzled me, since I thought that any hippy in good standing had to smoke a pack of marijuana a day, and I asked about it.

"You smoke mostly out of boredom," Chris said, which explained why I missed seeing them with any drugs. Fighting storms, landing in hayfields, learning formation and takeoffs and landings—boredom was not a problem that we had to face.

In the midst of my language practice I noticed the kids had begun to pick up flying jargon without any dictionaries at all.

"Hey man," I asked Joe one day, "this word 'rushing,' you know, I don't quite dig it. How do you, you know, use it in a sentence?"

"You can say, 'Man, I'm rushing.' It's the feeling you get on smoke when the top of your neck feels like it's going into the back of your head." He thought for a while, then brightened. "It's exactly like how you feel pulling out of a spin." I suddenly understood about rushing.

Words like "tail-dragger," "rag-wing," "touch-and-go," "loop," "hammerhead" popped up in their talk. They learned how to pull a propeller through by hand to start an engine, they followed us on the dual controls through every slip, skid, short-field landing, and soft-field takeoff we made. Even details, they picked up. Joe had his hands full flying formation one morning and called to me in the back seat of the Cub, "Could you give me a little up trim, please?" He didn't hear, but I had laughed at that. A week earlier, "trim up" had been something you did to a Christmas tree.

Then one evening around the fire, Chris said, "How much does an airplane cost? How much do you need to fly one for a year, say?"

"Twelve hundred, fifteen hundred dollars," Lou told him. "Fly it for two dollars an hour . . ."

Joe was astonished. *"Twelve hundred dollars!"* There was a long silence. "That's only six hundred apiece, Chris."

The fly-in at Oshkosh was a carnival that left them unimpressed. They had not been caught by airplanes as much as by the idea of flight itself, by the idea of riding some airborne motorcycle up off the ground, leaving roads and traffic lights behind, and setting out to discover America. More and more this began to occupy their thinking.

Rio, Wisconsin, was our first stop homeward. There we carried thirty passengers on joy rides over town. The kids helped the passengers into the planes, explained flying to those just come to watch, and found that it was quite possible for a fellow to break even this way, if he had a plane of his own. That afternoon we earned fifty-four dollars in contributions and donations, which bought us gas and oil and suppers for a few days to come. At Rio, the town treated us to a picnic complete with salads and hot dogs and beans and lemonade, balancing out those nights lost to wet bedrolls and hungry mosquitos.

Here Glenn and Michelle Norman left us to fly farther southeast, to meet friends and see farther into the USA.

"There's nothing more poetic or joyful-sad," Chris wrote in the journal, "than seeing a friend fly away in an airplane."

South we flew, four of us now in two airplanes, south and east and north again.

For crowded skies, that Monday afternoon, we saw a total of two other airplanes in all Chicago's metropolitan airspace.

For 1984, we saw the horses and buggies of the Indiana Amish on the country roads below, and three-horse teams hitched to plows in fields.

Our last evening out we landed in the hayfield of Mr. Roy Newton, not far from Perry Center, New York. We talked with him for a while, asking his permission to stay the night on his land.

"Course you can stay here," he said. "Except you won't light any fires, will you? The straw around . . ."

"No fires, Mr. Newton," we promised. "Thanks very much for letting us stay."

Later, it was Chris who spoke. "You can sure get away with murder, in an airplane."

"Murder, Chris?"

"Suppose we had come along with a car, or on bikes, or walking. What are the chances he'd be so nice about letting us stay here like this? But with airplanes, because it's getting dark, we're welcome to land!"

It didn't sound fair, but that's just the way it is. That is a privilege one has, as a pilot, and it was not lost on the kids.

Next day we were landed back to Sussex Airport, New Jersey, and the Invitational Cross-Country Adventure was officially finished. Ten days, two thousand miles, thirty hours of flying.

"I'm sad," Joe said. "It's all over. It was great and now it's all over."

It wasn't till late night that I opened the journal once more, and noticed that Chris Kask had made one last entry in it.

"I learned a *tremendous amount,*" he had written. "This has opened my mind to a whole bunch of things that exist outside of Hicksville, L.I. I've got a new perspective on things. I'm able to stand back from everything a little more and see it from a different viewpoint. Something I've felt on this is that it's an important thing not only to me but to everyone along and to everyone we met, and I realized this while it was happening, which is a very heady feeling. It caused many tangible and many intangible changes in my mind and emotions. Thanks."

There was my answer, there's what we can say to the kids who say "Peace" instead of "Hello." We can say "Freedom," and by the grace of a secondhand rag-wing lightplane, we can show them what we mean.

Too many dumb pilots

"It's not that there's too many pilots flying," somebody wise once said, "it's that there's too many dumb pilots flying!"

Lives an aviator who does not agree? Many as the leaves of the forest are the times I've flown into a traffic pattern exactly on altitude, just the perfect distance from the runway on downwind leg—just precise gliding distance to the field if the engine should fail, all nicely set up to turn to base leg. And I've looked out and seen, by Ned, some clod droning along a monster two-mile final approach, dragging his airplane to the threshold on sheer power, never considering remotely possible the chance that his fan might cease turning.

And there had gone my lovely pattern, as I gronked the throttle back, nosed up into slow-flight to save what I could of it. I have expressed more than once to my instrument panel that I behold a man with head of solid pine, down there, one not caring that when he flies a thoughtless pattern, he destroys everyone else's pattern, too, as each struggles to take spacing on him. I, gentle I, who never whisper at demonic stupidity about me on highways have spoken ill in the air of a fellow pilot. Why should that be?

I speak ill, perhaps, because I can expect occasional ignorance of anyone who crawls along the surface of the earth, but only perfection of anyone who chooses to lift into the sky, and it is a crushing disappointment to find otherwise.

Too many dumb pilots? Yes, indeed. Why, if everyone could be as good an aviator as I am, and as you, there would be no conflict, today, in general aviation, or questions about its future.

The answer is education. Educate that clod to fly a proper pattern through the simple instructional technique of chopping his power on that wallowing final approach—that'll teach him! Build new engines factory-guaranteed to fail at least once every five hundred hours, and we'd have good pilots everywhere in the sky.

So I mutter and fulminate and lecture my instrument panel, noticing where the offenders land (bouncing their touchdowns, of course), watching them in quiet wrath on the ground. Yet they are healed as soon as they are out of their aircraft, they become normal human beings, affable, kind, smiling, not the least aware of the havoc they have wrought upon my magnificent landing patterns. I watch, and finally shake my head, keep my silence, and walk away.

Then came one time, however, once when I bounced a landing. Me . . . Bounced.

Although no one was watching, although I would of course never repeat the offense, it was disquieting.

Disquiet magnified in the little town of Mount Ayr, Iowa, just at sunset over a narrow grass strip, attended only by sparrows and a meadowlark.

Three other airplanes flew with me, airplanes piloted by 1) a commercial charter pilot, 2) an airline captain on holiday, and 3) a third-year college student at the wheel of the first airplane he had ever owned.

It was getting dusky on the ground, and I worried about the kid. I turned down to land, and for some reason I had the devil's own time controlling the roll-out—I was all over the cockpit holding the biplane straight, and at that she used every foot of the strip. The airline captain came in next, and he landed hot and long, too. Then the charter pilot touched, and as conditions would have it, his landing was as bad as ours had been. By now I was quite concerned for the youngster . . . this was no easy challenge, coming down here, but the poor kid had to do it or be caught up in the night. We three on the ground got out of our machines and met in a little knot of worry.

"Spence, that's a tough one," I said toward the airline captain. "Think young Stu can pull it off?"

"Dunno. There's a wicked downdraft there at the end of

the strip . . ." We all wrinkled our foreheads and watched.

Stu didn't come in at once. He made one low pass over the grass, and then he did an odd thing: he turned around and landed in the opposite direction. Pretty as an Amendola painting . . . his airplane touched down three-point, rolled a few hundred feet, and stopped. We fell silent, the three of us.

In that silence, the youth shut down his engine and climbed down from his aircraft.

"What's the matter with you guys?" He said it in the respectless tone of the young and inexperienced. "How come you landed downwind? Is it me? A guy is supposed to land into the wind, isn't he?"

It was silent still, and he spoke again. "Dick? Spence? John? Why did you land downwind?"

It fell to me to be spokesman for the experienced pilots, for we three who had together logged something over fifteen thousand hours.

"well stu it's like this we landed downwind . . . ah . . . we landed downwind because we didn't want the sun to get in our eyes. flicker vertigo you know when that sun gets in your eyes through the prop . . ." I said this low and quick, hoping one of the others would jump in fast and change the subject.

"What do you mean?" Stu said, perplexed. "The sun's just set: the sun's been behind the hill for ten minutes! Say, you guys . . . you didn't . . . you didn't land downwind by mistake, did you . . . not by mistake?"

"ok well yes stu i was leading and i landed downwind if you must know by mistake and spence and john followed and did what i did. that's what happened. i'm hungry boy it's been a long day hasn't it spence. sure could use a bite to eat, don't you think there stu sure let's walk down the road and find us a supper . . ."

"BY MISTAKE! There's the windsock! All three of you, all these fantastic pilots . . . LANDED DOWNWIND BY MISTAKE!" The kids today I think are taught to rub things like that in.

He started laughing, then, and strangled it off only when our sullen glare made it clear that we did not find the matter funny, and that he would be thrown inverted into the river if there was not respect for elders forthcoming pronto.

And that is about the end of the story. Every once in a

while, like the graybeards of forty years flying who land occasionally at the wrong airport, it is our own head that is carved from knotty pine . . . that dumb clod in the sky is us!

What's to be done when a pilot as good as you and I has an occasional moment of lapse?

The answer's unchanged. Education. But the special education this time is that no matter how many times we have landed or taken a machine up into the air, we can never afford to do it half awake, or by habit. That with familiarity must come the knowing that the better we get, the more piercing crushing intolerable becomes stupidity when it is found in us.

This is called learning. Not one of the three of us oldtimers has landed downwind in the two years since, and there is some chance that we never will again. And we solemnly guarantee, as our promise and service to aviation, that the first time that kid Stu lands downwind, he will never once in all the rest of his life be allowed to forget it.

Think black

Think black. Think it above and below and all around you. Not a pitch black, but just a darkness without horizon or moon to give it reference or light.

Think red. Put some softly in front of you, on the instrument panel. Let it barely show twenty-two instrument faces with ghostly needles pointing to dim markings. Let the red flood gently around to your left and right. If you look, you can just see your left hand on the thick throttle, and your right holding the button-studded grip of the control stick.

But don't look inside, look out and to the right. Ten feet from the plexiglass that keeps pressure around you is a spot of red light, flashing.

It's attached to the left wingtip of the lead airplane in the formation. You know that the plane is an F-86F; that its wings are swept to a thirty-five-degree angle; that in its fuselage is a J47-GE-27 axial-flow jet engine, six fifty-caliber machine guns, a cockpit like yours, and a man. But you take it all on faith; you see only a dim red light, flashing.

Think sound. A dynamo's whine behind you, eerie, low and unceasing. Somewhere on the dim panel in front, an instrument is telling you that the engine is putting out ninety-five percent of its rpm; that fuel is being fed to it at a pressure of two hundred pounds per square inch; that there is thirty pounds of oil pressure at the bearings; that the temperature in the tailpipe, behind the combustion chambers and the spinning turbine wheel, is five hundred seventy degrees Centigrade. You hear the whine.

Think sound. Think the hiss of light static in the foam-rubber earphones of your crash helmet. Static that three other men in a sixty-foot radius are hearing. A sixty-foot radius at thirty-six thousand feet, four men alone-together swishing through the thin black air.

Push with your left thumb and four men can hear you talk, can hear how you feel, seven miles above a ground you cannot see. Dark ground, buried under miles of dark air. But you don't talk, and neither do they. Four men alone with their thoughts, flying on the flashing light of the leader's airplane.

Everything else about your life is normal, and everyday common. You go to the supermarket; the gas station; you say, "Let's eat out tonight!" But every once in a while you are far away from that world. In the high blackness of a star-studded sky.

"Checkmate, oxygen check."

You slide your plane out a little from the flashing light and look into the dim red of your cockpit. Hiding in a corner is a luminous needle, pointing two-fifty. Now your thumb hits the microphone button, there's reason to talk.

Your own words sound strange in your ears after the long quiet. "Checkmate Two, oxygen normal, two-fifty."

Other voices in the black:

"Checkmate Three, oxygen normal, two-thirty."

"Checkmate Four, oxygen normal, two-thirty."

Silence pours back in, and you close again on the flashing red light.

What makes me different from the man behind me in the grocery line? you wonder. Maybe he thinks I'm different because I have the glory-filled job of a jet fighter pilot. He thinks of me in terms of gun-camera film in the newsreels, and a silver blur of speed at an air show. The film and the speed are just part of my job, as preparing the annual budget report is part of his. My job doesn't make me any different. Yet I know that I am different, because I have a chance that he doesn't. I can go places he will never see, unless he looks up into the stars.

Still, it isn't my being here that sets me apart from those who spend their lives on the ground, it's the effect this high, lonely place has on me. I get impressions that can't be equaled anywhere else, impressions that he'll never feel. Just to think of the reality of the space outside this cockpit is a strange feeling. Eleven inches to my right, eleven to my

left, is a place where man can't live, where he doesn't belong. We flick through it like frightened deer across an open meadow, knowing that to stop is to flirt with death.

You make tiny automatic motions with the stick, correcting to keep in position on the flashing light.

If this were day, we'd feel at home; a glance downward would show us mountains and lakes, highways and cities, familiar things we can glide down to and be at ease. But it isn't day. We swim through a black fluid which hides our home, our earth. Engine failure now, and there's no place to glide to, no decision to make where to go. My plane can glide for a hundred miles if the rpm falls to zero and the tailpipe cools, but I'm expected to pull the handgrips, squeeze the trigger, and float down through the darkness in my parachute. In the daylight, I'm expected to try to save the airplane, try to put it on a landing strip. But it's night, it's dark outside, and I can't see.

The engine whirls faithfully on, and stars shine steadily. You fly the flashing light, and wonder.

If Lead's engine failed now, what could I do to help him? Simple answer. Nothing. He flies now twenty feet away, but if he needed my help, I'd be as far away as Sirius, above. I can't take him in my cockpit or hold his airplane in the air, or even guide him to a lighted field. I could call his position to rescue parties, and I could say 'Good luck' before he fired his ejection seat into the black. We fly together, but are as alone as four stars in the sky.

You remember talking to a friend who had done just that, left his plane at night. His engine had been on fire, and the rest of the formation was completely powerless to help. As his plane slowed and started down, one of them had called, "Don't wait too long to get out." Those helpless words were the last he had heard before he fired into the night. Here was a man he had known and flown with, who had eaten dinner with him, who had laughed at the same jokes with him, saying, "Don't wait too long . . ."

Four men, flying alone together through the night.

"Checkmate, fuel check."

Once again, the voice from Lead cuts into the silence of the engine's airy roar. Once again you move away, read the dim needle, pointing.

"Checkmate Two, twenty-one hundred pounds," your stranger's voice calls into the thin static.

"Checkmate Three, twenty-two hundred."

"Checkmate Four, twenty-one hundred."

Back in you slide, back to the flash of the red light.

We took off just an hour ago, and already the fuel says it's time to go down. What the fuel says, we do. Strange what a complete respect we have for that fuel gage. Pilots who respect neither laws of man nor of God respect that fuel gage. There's no getting around its law, no hazy threat of punishment in the indefinite future. Nothing personal. "If you don't land soon," it says coldly, "your engine will stop while you're in the air, and you will bail out into the dark."

"Checkmate, descent check, and speed brakes . . . now."

Black air roars outside as the two metal slabs that are your speed brakes push into the slipstream. The red light keeps flashing, but now you push forward on the stick to follow it down, toward the invisible ground. Abstract thoughts fly to the depths of your mind, and you concentrate on flying formation in the steep descent. Those thoughts are for high places, for as the earth approaches, there is more to do to fly the airplane safely. Temporal, concrete, life-depending thoughts jumble your mind.

Move it out a little, you're too close to his wing. Fly smooth, don't let a little rough air bounce you out of formation.

Impersonal turbulence pounds your plane as you turn together toward the double row of white lights that mark the waiting runway.

"Checkmate, turning initial, three out with four."

"Roger Checkmate, you're number one in traffic, winds west northwest at four knots."

Funny, that in our sealed cockpits at three hundred miles per hour, we still must know about the wind, the ancient wind.

"Checkmate's on the break."

No thoughts now, but reflexes and habits as you land. Speed brakes and landing gear, flaps and throttle; you fly the landing pattern, and in a minute there is the reassuring squeak of wheels on concrete.

Think white. Think glaring, artificial light shining reflected from waxed tabletops in the flight shack. A sign on the blackboard: "Squadron Party . . . 2100 hours tonite. All the beer you can drink—FREE!"

You're down. You're home.

Found at Pharisee

It happened on a Tuesday, at Pharisee, Wyoming. I remember that I had just grounded myself for a week, because the approved mechanics were busy and wouldn't be able to change my airplane's oil till Tuesday next. I had logged twenty-four hours fifty-seven minutes since my last twenty-five-hour oil change, so of course I couldn't fly.

As I turned to leave the FAA-approved repair station, there was a great thunder in the sky, and a dozen lightplanes landed suddenly on the grass, where it was forbidden to land, and, I later learned, without radio. They converged upon the FAA-approved repair station like multiple lightnings, and a dozen masked, black-clad men leaped from the cockpits and surrounded us, .44s drawn and cocked.

"We'll take all your technical files, right now," the outlaw leader said in a low, calm voice. A black silk cape hung about him, and from the cool manner in which he pointed his revolver, it was plain that he had done this many times before. "Everything you have, anything written about any airplane, any engine, bring it out here, please."

It was preposterous, incredible, in this day . . . a holdup! I started to cry out, but the FAA-approved inspector, without moving a muscle, said, "Do as he says, boys; give 'em the tech data files."

Three of the approved mechanics backed toward the office, covered by the outlaws.

"What's going on here?" I asked. "What is this?"

"Quiet, over there."

"What do you mean, quiet? This is illegal! HELP! F.A.A.! OUTLAWS!"

When I woke up, I was lying on a cot in a cave of rock, well lit and apparently part of a vast complex, a hidden community. My airplane was chocked in a stone T-hangar cut opposite a huge sliding wall, and a black-clad outlaw had just changed the engine oil. He was removing a magneto contact-breaker assembly now, and it jarred me to action.

"Stop! You can't do that! You're not a licensed mechanic! Put that back on!"

"If I'm not a licensed mechanic, I can't put it back on, can I?" He spoke quietly, without looking at me. "Sorry we had to bring you along, but Pharisee had more data than we planned on taking, and we had to borrow your airplane to help carry the load. We figured you wouldn't want to be left behind. And your left magneto dropped fifty rpm on run-up."

You don't reason with people like that, but I was still confused, not thinking clearly. "What's the matter with fifty rpm? I can have a seventy-five rpm drop and still be legal."

"Sure you can, but being legal doesn't make it right." He was quiet for a moment. "Just like being right doesn't make it legal. This magneto misses fire every minute and a half in the air. You never noticed that?"

"How could I notice it? I never fly on one magneto in the air. I check them both before takeoff and if there is less than a seventy-five rpm drop . . ."

". . . you go ahead and fly."

"Of course I do. I learned by the book and I fly by the book." I have always been proud of that.

"Heaven help us," was the only reply from the outlaw.

A few minutes later, as he worked, I gathered courage, and spoke. "What are you going to do with me?"

"Let you go. As soon as we pay you for the use of your airplane. Price of replacing this breaker spring will just about cover it."

"Pay me? But you're outlaws! That's not a legal repair! Who will sign it off in the logbook?"

The black-clad desperado laughed, low in his throat. "That's your problem, friend. All that matters to us is that an airplane works as it was designed to work. The paperwork is up to you."

"What about all that technical data you took?" My words cut like razors. "Were you so noble as to pay for that?"

"Overpaid, if you ask me. But that's the way Drake has

to have it. We left a zero-since-major engine at Pharisee . . . tolerance to a ten-thousandth all the way through, all our own best work. Drake's personal guarantee for three thousand hours flying. Man. The things we give to get more technical information . . ."

"But if you outlaws overhauled it here, it has no logs, it hasn't been signed off!"

He laughed again, setting a timing disc on the propeller shaft. "You're right. It hasn't been signed off. We have left them the best-overhauled engine in the world today, and it isn't legal. They'll have to tear it down, won't they? . . . change the tolerances, break the guarantee. When they get it back together, it will be just another engine, with a fifty-hour warranty. But legal, friend, legal!"

He touched a set of buttons beneath a dial on the wall. "Looks as if you might be staying the night. Wind's twenty miles per hour on the north strip here. Twenty-three on the south."

The finality of his words frightened me. "There's nothing wrong with twenty miles per hour," I said. "That's less than half the stalling speed of my airplane, and according to the book, if the wind is less than . . ."

"That much wind in these mountains will blow you to pieces with what you know about your airplane."

"If you had taken the time to examine my logbook," I said icily, "you would have seen that . . ."

". . . that you have 2648 hours and 29 minutes total flying time. Our computers have analyzed the kind of flying you've done. A thousand of your hours are logged on autopilot, and the rest of it was spent trying to fly like one. You have the equivalent of sixteen hours and sixteen minutes real flying time, our kind of flying time. That's not enough to fly out of here safely in a twenty-mile breeze." He turned the propeller slightly.

"Now just a minute. I don't know what kind of screwy computer you have, but I know I can fly my own airplane."

"Sure you can. You have logged 2648 hours in your little book." He turned so suddenly I jumped, and his words slammed rapid-fire into the rock walls. "How much altitude do you lose in a one-eighty downwind turn, if the engine stops on takeoff? How long does it take your gear to extend on battery power only? What happens when you land with the wheels only partly down? How do you make a minimum-damaged forced landing? If you have to fly through power lines, where should you hit them?"

It was quiet for a long moment. "Well, you never turn back to the runway if the engine stops on takeoff; that's in the book . . ."

"And the book lies!" He was immediately sorry for his outburst. "Excuse me. Let's say that the engine stops on takeoff after you have climbed to five thousand feet and circled so that you were over the end of the runway?"

"Well, of course. I could turn . . ."

"One thousand feet?"

"That's plenty high enough to . . ."

"Five hundred feet? Three hundred feet? One hundred feet? Do you see what I mean? Our instructors teach that a pilot should know his turn-around altitude for every takeoff he makes."

"Then you have outlaw instructors, too."

"Yes."

"And I suppose they teach spins and lazy eights . . ."

". . . and prop-stop glides and forced landings to touchdown and aerobatics and flying without trim or flight controls and . . . and a lot of things you've never even thought about in your hours on autopilot."

I replied with penetrating sarcasm. "Your students, I suppose, all get their licenses in the minimum thirty-five hours?"

"Our students never get a license. We're outlaws here, remember? We judge our ability by how well we know ourselves and our airplanes, day to day. We leave the paperwork and the licenses to people who live by regulations instead of knowing." He finished with the magneto, and removed the timing disc. "Let's go eat."

The dining hall was a gigantic underground cavern, lighted by tall bright panels printed with diagrams and cutaways of engine and airplane components. The hall was half filled with black-clad men, and rows of black hats and black gunbelts hung from the black hat-racks. I noticed, with a shock, that a black silk cape hung from the first rack.

"Drake would like the pleasure of your company."

The last thing I wanted then was to sup with the leader of this outlaw band, but I dared not say so. I followed my guide to a corner table, at which sat a lean, square-jawed figure, clothed all in black.

"Here he is, Drake. We made up a new breaker spring for his left mag, and our debt to him is paid."

"Thank you, Bart." The voice was low and sure, obviously the voice of a madman, and a man to be treated as such.

"I demand my rights," I said firmly. "I insist that you release me at once, and allow me to leave this robbers' roost."

"You have your rights," he said, "and you may leave whenever you wish. You know, of course, that at the moment our downdrafts exceed your ability to make your airplane climb. We've also found that your number four connecting rod is cracked, and may fail at any time. If it fails within fifty miles of this room, you do not have the ability to land your airplane without destroying it. Knowing these things, if you still wish to leave, you may leave. You might be lucky in the wind, and the rod might not break immediately."

He was obviously a crazed gunman, and I destroyed his point at once. "I have flown over fifteen hundred hours in that very airplane, Mr. Drake, and I certainly ought to be able to fly it safely in this paltry wind. And if you had not been so hasty to kidnap me, you would have seen that my engine has only fifty hours since major overhaul by a reputable firm, for which I paid $1,750, and for which I have a receipt and an inspector's signature in the logbook."

The meal was silently served, and during the serving Drake looked at me with the hopeless, slightly sad look of a craven criminal.

"The connecting rod of your number four cylinders does not even know what a logbook is. Will it comfort you to read your logbook and trace the inspector's signature when your propeller stops turning and there is no place for you to land?"

The man, I had to admit, was uncanny. Actually if such an impossible thing as a fifty-hour engine quitting in flight could happen, it *would* be a comfort to read the inspector's name again, but the way he said it made the idea of depending on a signature sound silly. I set him straight.

"One chance in a million, my dear Drake, and I am not that one. As long as a pilot is legal, he is safe. Furthermore, anything that breaks the regulations of the Federal Aviation Agency is unsafe. Surely a government agency should know what is safe and what is not." To my amazement, the madman laughed. Not scornfully, but as though he had thought of something that he found humorous.

"You are priceless," he said, laughing still. "Or perhaps I misunderstand. When you speak of this infallible government agency, do you mean the same government agency that removed training in spins from its pilot requirements? The same agency that now says it is well merely to teach approaches to stalls rather than full stalls, when the stall-spin is a major cause of modern pilot fatality? Do you mean the same regulating body that sets a brand-new mechanic to work on an old-style radial engine, while it brands 'outlaw' the unlicensed owner who knows more about the engine than the mechanic will ever learn? The same agency that requires itself to hire ten blind paper shufflers for every able man it has?"

He laughed again, setting down his fork. "The same agency I wrote so long ago for information, that told me, 'It is not considered essential to flight safety for a person to know the actual design load factor of his airplane,' and refused to send me information from public files?"

"I mean the Federal Aviation Agency," I said, and I spoke with solemn dignity. The brigands round about clearly had no respect for authority, for they looked at me and smiled, as though they could hear what I said and as though they too had thought of something humorous. I decided then to destroy their leader's position in front of them all, and raised my voice so that all could hear.

"Then you feel that the Federal Aviation Agency is all bad, Mr. Drake, and should be abolished?"

"Of course not," he said quietly. "Some kinds of aviation, airlines, for instance, need central coordination to fly efficiently, to serve their customers and the country."

"Well, if you don't think it should be abolished, why are you not a law-abiding man, a follower of regulations?" I had destroyed the man by his own logic, and I had to smile. I awaited his abject concession.

"Just because I say I enjoy a steak from time to time, my friend, does not mean I want a cow stuffed down my throat. We outlaws fly and maintain our own airplanes for fun, we don't fly DC-8s on international airways."

Curse him. "The rules, man, the rules! They are made by the FAA for our own safety!"

"Ah, my honored guest," the outlaw said, and leaned forward across the table, "you seek your god in rule books and idols made by men, and all the while that god is within you.

Safety is that which you know, not what somebody else thinks it would be nice for you to comply with. Ask your FAA agent for the approved definition of safety. There is none. How can any agency guide toward that which it can't even define?"

"You poor lonely outlaws," I said, with as much pity as I could feign for the lunatic. "There are so few of you . . ."

"Think so?" my captor said. "Open your eyes. In the cities, with hard-surface runways and FAA offices crowded in the terminals, we are few. But come away from your executive transportation centers someday and see what is going on in the other ninety-nine percent of the country. Outlaws. It is not only impossible to fly daily without breaking Federal Air Regulations, but following them blindly can kill a man."

"An empty slogan, my good fellow."

"Is it? Fly in two-mile visibility into a controlled airport, sometime, with no radio. It isn't legal to land, is it? If you are seen landing and the FAA isn't in the mood to overlook the law that day, a violation will be filed against you.

"So you keep flying, hoping for an uncontrolled airport nearby. The weather goes bad around you, but you've never landed in a pasture—that sort of thing is considered dangerous and is not in the flight training requirements. It is now raining hard and you can't find an airport, so you decide that with your five hours of instrument hood training you are able to climb through the cloud to on top in uncontrolled airspace. What is instrument training for, if not for use in an emergency? By evoking the emergency prerogative section of the General Operating Rules, you can even do this legally. But your chances of coming through alive are zero.

"Just one instance," he said, "one logical everyday instance in which blind obedience to law will kill you. Want more? Plenty more examples, and lots and lots of outlaws. We're content to let the FAA live in its little dreamworld, as long as it doesn't make us try to live there, too. And it doesn't. I used to be an editor for an aviation magazine, and I had the chance to talk to many an official agent of the FAA. I found that the experienced men agreed with the outlaws right down the line, as long as I promised not to quote them on it. One of them said, 'There are more outlaws in the FAA than out of it!' Word for word, my friend, from a high-ranking regional official of your agency."

At my command, the man obediently passed the salt. "There are quite a few old-time pilots in the FAA who know us well," he went on, "and who know that our kind of safety works better than the official kind, and so don't apply the law to us, or bend it sharply for us. We've all agreed to be very quiet about the fact that a great number of regulations are ridiculous violations of common sense, and we've agreed that no one will rock the boat. We're grateful that the old-timers are there, of course. If anyone seriously tried to enforce the regulations on maintenance, for instance, virtually every owner of a low-cost lightplane would have a price on his head, and would have to counter-attack for his very survival as an airplane owner. The magnitude of that counterattack would destroy a great many people in the FAA, and it would reform the law. The end result would be good, certainly, but the process would be so painful that none of us has the courage to begin it. We are happy as long as we're left alone. The FAA is happy as long as no one shatters its dreamworld about the law-abiding little guy."

My patience came to an end, I had had enough of this self-righteous prattle. "Admit it, Drake," I said. "You're looking for a license to fly recklessly, to do whatever you wish, whether it be safe or not. You don't care whether you live or die, but how about the innocent people on the ground who are snuffed out when your reckless nonsense pays you in full?"

He laughed. "My friend, you do a lot of flying at night, don't you?"

"Of course I do. An airplane is for transportation, day and night. What's that to do with your recklessness?"

"Do you wear a parachute when you fly at night?"

"Of course not. What a juvenile thought!"

"What do you do, then, if your engine stops at night?"

"I have never had an engine failure in flight, Mr. Drake, and I do not intend to have one."

"Isn't that interesting!" He was silent for a moment, studying the engine diagram woven into the tablecloth. "There is not an outlaw here who would fly an airplane at night without a parachute, unless the moon was so bright that he constantly had a landing place in sight. We don't believe that engine failures never happen, and if we can't see to land, and if we can't carry a parachute, we don't fly. There's not a pilot here, except yourself, who would fly over an undercast of fog, or over a ceiling lower than he can shoot a forced landing from.

"Yet no-parachute night flying is perfectly legal, and flying on top of any amount of fog is FAA-approved. Our rule says that pure safety is pure knowledge and pure control. Whether our airplane has one engine or two is immaterial. If we can't see to land, and if we can't carry a parachute, we don't fly."

Naturally, I didn't listen to a word the man said. The only safety that wildman would ever know would be the safety of a prison cell.

"Your connecting rod," he went on "is legal right now. It is FAA-approved and it is all signed off. But it is cracked and it is going to break soon. If you had the choice, would you rather have the crack in the rod or that signature in your log?"

I could only be firm with him. "Sir, the mechanic and the inspector are responsible for their work. I am entirely within my rights to fly that airplane exactly as it is."

He laughed once more, a curiously friendly sound, as though he meant me no harm. At that moment I knew I would escape his lair, and soon.

"All right," he said, not knowing my thoughts. "The inspector is responsible, and you are innocent. All you have to do is let your airplane be destroyed in these mountains because you are not required to know how to survive in any land that you fly over. Everyone else is responsible, you are just the guy who does the dying. Is that it?"

That is it, of course, but again he made it sound foolish and wrong. But who can believe a band of outlaws, living in the badlands, flying and maintaining their airplanes without licenses just because they happen to know how an engine works or how an airplane flies? Radicals and extremists all, and there should be a law against them. Well, of course, there is a law.

Outlaws is what they are, and when I return to a law-abiding city, I'll see that the FAA files serious charges against them all, and revokes their . . . and comes out here and puts them in prison. They think they're so much better than everyone else, just because they know how to hold a wrench and land without power. But do they know about approach control? What do they do in the traffic pattern if the tower doesn't give them permission to land? They'd sing a different tune, then, and I'd reach over when they beg me to save them and I'd ask the tower, "Respectfully begging your permission to land," and then I won't have to know my airplane or how it flies because the tower has cleared me number one.

I abruptly took leave of Drake and his unsavory fellows, and neither he nor his men made any move to stop me. They no doubt saw my anger, and thought it much safer to hold their peace in my presence.

Back in the rock hangar, I found the button that slid the wall away, and since the outlaws were now clearly afraid of a law-abiding man, I took time to write this all down, every word we said, to use as evidence in the FAA hearings that will send these men to prison. Those wonderful, simple hearings, in which the FAA, because it knows what is best for us, can both prosecute us and judge us fairly. Fortunately, these wild ones are surely the only men of their kind in the country.

Note to myself: Type all notes following, since ruf air makes pencil words hard for prosecutor to read. Wouldn't have thot wind 20 so ruf. Save this paper, tho, show outlaws they wrong. Can fly out of their mountains with one hand, make notes with other.

Downdrafts bad. 1500 fpm down, tho full power and climb speed. Must hit updraft soon.

There. Worst is behind, and outlaws soon to justice. I see Pharisee airport, and I could almost stretch glide from here unless—chance in a million . . . chance in billion—the engine qui

School for perfection

I had flown west for a long time. West through the night, then south, then sort of southwest, I guess, not caring. You don't care too much about maps and headings when you've just lost a student. You go off by yourself, after midnight, and think about it. It had been an unavoidable accident; one of those rare times when fog forms right out of mid-air and in five minutes the visibility goes from ten miles down to zero. There had been no airport nearby; he couldn't land. Unavoidable. By sunup, the country around me was strange and mountainous. I must have flown quite a bit farther than I thought, and the fuel gage pointers were both bouncing on E. Lost, with the sun barely up, it was pure luck that I saw a green-painted Piper Cub rocking its wings to me, turning to land on a tiny grass strip at the base of a mountain. It touched the ground, rolled for a moment, then abruptly disappeared into a wall of solid rock. The place was empty and still as a frontier wilderness, and for a moment I thought that I had imagined the Cub.

Still, that little strip was the only possible place to land an airplane. I was glad I had taken one of the 150s, instead of the big Comanche or the Bonanza. I dragged up to the field, full flaps and power, facing right into that granite wall. It was the shortest landing I could make, but it wasn't short enough. Power off, flaps up, brakes on, we were still rolling at twenty knots when I knew we were going to hit the wall. But there was no impact. The wall disappeared, and the 150 rolled to a stop inside a huge stone cavern. It must have been a mile long, that place, with a great long runway. Airplanes of all types and sizes were parked about, each painted in dappled green camouflage. The Cub that had landed was just shutting down its engine, and a tall, black-clad fellow stepped from the front seat and motioned me to park alongside.

Under the circumstances, I could only do as he asked. As I stopped, another figure emerged from the back seat of the Cub. That one was dressed in gray; he couldn't have been more than eighteen years old, and he watched me with mild disapproval.

When my engine stopped, the man in black spoke in a low, even tone that could only have been the voice of an airline captain. "It must not be much fun, losing a student," he said, "but it shouldn't make you forget your own flying. We had to make three passes in front of you before you finally saw us." He turned to the youth. "Did you watch his landing, Mr. O'Neill?"

The boy stiffened. "Yes, sir. About four knots fast, touched down seventy feet long, six feet left of centerline . . ."

"We'll analyze later. Meet me in the projection room in an hour."

The youngster stiffened again, inclined his head slightly, and left.

The man escorted me to an elevator and pressed a button marked Level Seven. "Drake's been wanting to see you for some time," he said, "but you haven't been quite ready to meet him until now."

"Drake? You mean Drake the . . ."

He smiled, in spite of himself. "Of course," he said, "Drake the Outlaw."

In a moment, the door hissed open, and we walked a long, wide passageway, carpeted and quiet, tastefully decorated in detail diagrams and paintings of aircraft in flight.

So he really exists, I thought. So there really is such a man as the Outlaw. When you operate a flying school, you hear all kinds of strange things, and from here and there, I had heard of this man Drake and his band of flyers. For these people, the story went, flight had become a true and deep religion, and their god was the sky itself. For them, it was said, nothing mattered but reaching out and touching the perfection that is the sky. But the only evidence of Drake's existence was a few handwritten pages, an account of meeting the man, found in the wreck of an airplane that did not survive a forced landing. It had been printed once in a magazine, as a curiosity, and then forgotten.

We entered a wide, paneled room, so simply furnished that it was elegant. There was an original Amendola painting of a C3R Stearman framed on one wall; on the other was a fine-detail cutaway of an A-65 engine. My guide disappeared, and I couldn't help but examine the C3R. There was no flaw anywhere in it. The fasteners were there on the cowl, the rib-stitching of the wings, the reflections in the polished fabric. The Stearman fairly vibrated on the wall, caught in the instant of flare, just above the grass.

If only reality could be as perfect as that painting, I thought. I had been to so many seminars, heard so many panel discussions affirm in parrot voices, "We're only human, after all. We can never be perfect . . ."

For a second I wished that this Drake could live up to his legend, say some magic word, tell me . . .

"We can be perfect, my friend."

He was about six feet tall, dressed in black, with the lean, angled face that independence gives to men. He could have been forty years old or sixty, it was impossible to tell.

"The Outlaw himself," I said, surprised. "And you read minds, as well as fly airplanes."

"Not at all. But I think you might be tired of excuses for failure. Failure," he said, "has no excuse."

It was as if I had been climbing up through clouds all my life, and in this moment had broken out on top. If he could only back up those words.

Yet suddenly I was very tired, and threw the full weight of my depression at him. "I'd like to believe in your perfection, Drake. But until you show me the perfect flight school, the perfect staff of instructors, with no failures and no excuses, I can't believe a word you say."

It was my last hope in the world, a test for this leader of these very special outlaws. If he was silent now, if he apologized for his words, I'd sell my flying school cold, take the Super Cub back to Nicaragua for a living.

Drake's answer was a half-second smile. "Follow me," he said.

He led the way into a long hall, lined in glowing aviation art and pedestals mounting bits and pieces of world-famous airplanes. Then we turned down a narrow corridor and abruptly into cool air and sunlight, at the brink of a steep, grassy slope. The grass fell away some fifty feet, and where it merged with level ground was a huge fluffy square of what looked like feathers, a hundred yards on a side and perhaps ten feet deep.

A man, gray-haired, dressed in black, stood by the feather pile and called up the slope. "All right, Mister Terrell, whenever you're ready. No hurry. Take your time."

Mister Terrell was a boy of fourteen or so, and he stood to our left, on the edge of the slope. Resting on his shoulders was a great frail set of snow-linen wings, thirty feet from tip to tip and casting a transparent shadow on the grass. He took a breath in readiness, reached forward, and gripped the adhesive-taped bar of the main wing beam. Then all at once he ran forward, tilted the wings upward, and lifted free of the hillside. He flew for perhaps twelve seconds, swinging his body as a gymnast would, in slow feet-together motions that balanced the white wings smoothly down through the air.

At no time was he more than ten feet above the slope, and he dropped free of the wings a second before his feet touched the feathers. It was all slow and graceful and free, a kind of dream turned into white linen and green grass.

Voices drifted up, tiny, from the meadow. "Just sit there for a while, Stan. Take your time. Remember what it felt like. Remember it through, and when you're ready, we'll take the wings up and fly again."

"I'm ready now, sir."

"No. Live it through again. You're at the top of the hill. You reach up to the spar. You run forward three steps . . ."

Drake turned and led the way into another long corridor, into a different part of his domain. "You ask about a flight school," he said. "Young Mister Terrell is just beginning to fly, but he has spent a year and a half studying the wind and the sky, and the dynamics of unpowered flight. He has built forty gliders. Wingspans from eight inches up to the one you just saw—thirty-one feet. He made his own wind tunnel and he has worked with the full-size tunnel on Level Three."

"At that rate," I said, "it's going to take him a lifetime to learn to fly."

Drake looked at me, and raised his eyebrows. "Of course it will," he said.

We turned now and then, through a maze of halls and corridors. "Most of the students choose to spend about ten hours a day around the airplanes. The rest of the time they give to other work, their own studies. Terrell is building an engine of his own design, for instance, learning casting and machining down in the shops."

"Oh come on," I said. "This is all very nice, but it's just not . . ."

"Practical?" Drake said. "Were you going to say that it isn't practical? Think, before you say it. Think that the most practical way to bring a pilot to perfection is to reach him when he is caught with the idea of pure flight, before he decides that a pilot is a systems operator, pushing buttons and pulling levers that keep some strange machine in the air."

"But . . . bird wings . . ."

"Without the bird wings, there can be no perfection. Imagine a pilot who has not only studied Otto Lilienthal but who has been Otto Lilienthal, holding his bird wings and leaping from his hill. Then imagine the same pilot, not only studying the Wrights, but building and flying his own powered biplane glider; a pilot who keeps within him the same spark that fired Orville and Wilbur at Kitty Hawk. After a while, he might be a pretty good pilot, don't you think?"

"Then you are taking your students, firsthand, through the whole . . . history . . ."

"Exactly," he said. "And the next step from the Wrights might be . . . ?" he waited for me to fill in the blank.

"A . . . a . . . Jenny?"

We turned the corridor into the sunlight again, at the edge of a broad, flat field, furrowed with the mark of many tailskids. A JN-4 teetered there, painted olive drab and camouflaged as the airplanes in the main cavern had been. The OX5 engine pushed a big wooden propeller around with the sound of a giant, gentle sewing machine whisking a needle through deep velvet.

A black-clad instructor stood by the rear cockpit.

"She'll fly a little lighter, Mister Blaine," he said, over the sewing-machine sound, "and she'll lift off a little quicker, without my weight. Three landings, then bring her back here."

In a moment, the Jenny was trundling out into the wind, moving faster, tailskid lifting just clear of the grass and holding there, at last the whole delicate machine rising slowly, so that I could see pure sky under its wheels.

The instructor joined us, and inclined his head in that curious salute. "Drake," he said.

"Yes, sir," Drake said. "Young Tom doing all right?"

"Quite all right. Tom is a good pilot—might even be an instructor, one day."

I could restrain myself no longer. "The boy's a bit young for that old airplane, isn't he? I mean, what if the engine stops now?"

The instructor looked at me, puzzled. "Pardon me? I don't understand your question."

"If the engine stops!" I said. "That's an old engine! It can quit in flight, you know."

"Well of course it can quit!" The man looked to Drake, as if he wasn't sure that I was real.

The outlaw leader spoke patiently, explaining. "Tom Blaine overhauled that OX5 himself, he machined parts for it. He can diagram the engine blindfolded. He knows where it's weak, he knows what kind of failures to expect. But most of all, he knows about forced landings. He began to learn forced landings with his first glide down Lilienthal Hill."

It was as if a light had been turned on; I was beginning to understand. "And from here," I said slowly, "your students go on into barnstorming and racing and military flying, right on through the history of flight."

"Exactly. Along the way, they fly gliders, sailplanes, homebuilts, seaplanes, dusters, helicopters, fighters, transports, turboprops, pure jets. When they're ready, they go out into the world—any kind of flying you can name. Then, when they've finished flying on the outside, they can choose to return here as instructors. They take one student, and begin to pass along what they've learned."

"One student!" I had to laugh. "Drake, it's clear that you've never had to operate a flight school under pressure, where the stakes are high!"

"In your flying school," he said mildly, "what are the stakes?"

"Survival! If I don't keep turning out pilots and bringing in new students, I'm through, I'm out of business!"

"Our stakes are a little different," he said. "It's up to us to keep flight alive in a world of airplane-drivers—the people who come out of your school, concerned only with moving straight and level from airport to airport. We're trying to keep a few real pilots left in the air. There's not too many left who don't carry that book of excuses, those 'Twelve Golden Rules,' next to their heart."

I couldn't have heard him right. Was Drake attacking the Golden Rules, distilled from so much experience?

"Your Golden Rules are all don'ts and nevers," he said, knowing my thought. "Ninety percent of the accidents happen in these conditions, so you must avoid the conditions.

The one last logical step they didn't print is 'One hundred percent of the accidents are caused by flying, so for complete safety, you must stay on the ground.' It was Golden Rule number eight, by the way, that killed your student."

I was thunderstruck. "It was an unavoidable accident! The temperature-dewpoint came together without being forecast, the fog just formed around him in five minutes. He couldn't reach an airport!"

"And rule eight told him never to land away from an airport. In his last five minutes of visibility, he flew over eight hundred thirty-seven landing places—smooth fields and level pastures—but they were not 'designated airports, with known current runway maintenance' so he didn't even think about landing, did he?"

It was quiet for a long time. "No," I said, "he didn't."

We were back in his office before he spoke again.

"We have two things here that you don't have in your school. We have perfection. We have time."

"And machine shops. And bird wings . . ."

"All the effects of time, my friend. The live history, the motivated students, the instructors . . . they're all here because we decided to take time to give a pilot skill and understanding, instead of listing rules.

"You talk about your 'crisis in flight instructing' on the outside, you're going through a frenzy of renewing all your instructors' licenses. But every bit of it is wasted unless the instructor is given time with his student. A man learns to fly on the ground, remember. He just puts that learning into practice when he steps into an airplane."

"But the tricks, the bits of experience . . ."

"Certainly. Prop-stopped forced landings, downwind takeoffs, control-jam flying, zero-G stalls, full-blackout night landings, off-airport landings, low-level cross-countries, formation flying, pride, instrument flying and no-instrument flying, low-altitude turn-arounds, flat turns, spins, skill. None of it taught. Not because your instructors don't know how to fly, but because they don't have time to teach it all. You think it's more important to have that scrap of paper, that flying license, than it is to know your airplane. We don't agree."

I threw the last of my resistance at him, as hard as I could. "Drake, you live in a cave, you've got nothing to do with reality. I can only pay my instructors for the hours they fly, and they can't afford to spend nonflying time talking with their students on the ground. If I'm going to survive, I've got to keep my planes and instructors in the air. We've got to put the students right through, give 'em forty hours and a copy of the 'Twelve Golden Rules,' get 'em ready for the flight check, and then start all over again with the next bunch. In a system like this, you're bound to have accidents now and then!"

I listened to myself, and all at once I was filled with loathing. It wasn't somebody else saying those words, fighting to defend failures, that was me, that was my own voice. My student's death wasn't unavoidable; I had murdered him.

Drake said not a word. It was as if he had refused to hear me. He lifted a tiny glider from his desk and launched it carefully into the air. It turned one full circle to the left and slid to a stop precisely in the center of a small white X painted on the floor.

"You might be just about ready to admit," he said at last, "that if your system involves accidents, then the solution is not to find excuses for the accidents. The solution," he said, "is to change the system."

I stayed a week at the cave, and I saw that Drake had not missed a single avenue that would bring perfection in flight. Instructors and students held a very formal relation, on the ground, in the air, in the shops and special-study areas. An incredible respect for the men and women who were instructors, almost a worship of them, filled Drake's domain. Drake himself called his instructors "sir," and the flying records of each one of them was printed and open to the students.

Sunday afternoon was a four-hour air show, with formation flying demonstrations of student-built airplanes, and a low-level aerobatic show by one of the best-known air-show pilots in the Southwest. Drake's influence and ideas ran deeper than I had dreamed . . . I began to wonder about a few other excellent pilots I knew; ag pilots, mountain pilots, airline captains who flew sport planes in their spare time. Could it be that they had some tie with Drake, with this school?

I asked, but Drake was enigmatic. "When you believe in something as true as the sky," he said, "you're bound to find a few friends."

The man operates an incredible flying school, and when

it was time to leave, I frankly told him so. But one thought persisted. "How can you afford it, Drake? This didn't all come out of thin air. Where do you get your money?"

"The students pay for their training," he said, as if that explained everything.

I must have stared at him rather dumbly.

"Oh. Not at the start. Not one student ever had a penny to his name, at the start. They just wanted to fly, more than anything else in the world. But every student pays what he thinks his training has been worth. Most give about ten percent of their income to the school, as long as they live. Some give more, some less. It averages about ten percent.

"And ten percent from a thousand bush pilots, a thousand military pilots, a thousand airline captains . . . it keeps us in gas and oil." Again, that half-second smile flashed across his face. "And it keeps them in the knowledge that there will be other pilots coming along that know more about flying than how to steer an airplane."

Heading north and east, flying back onto my map, I couldn't get his words out of my mind. To teach more about flying than how to steer an airplane; to take time with the students; to offer them the priceless thing that is the ability to fly.

I can change my school, I thought. I can choose my students carefully, instead of taking everyone that walks in the door. I can ask them to pay what the instruction is worth. I can pay my instructors four times what I'm paying them now; make instruction a profession instead of an odd job. Some extra training aids, perhaps—an engine dismantled, a cutaway airframe. My instructors' experience written for their students to read. Pride. Some firsthand history, some aerobatics, some soaring. Skill. Not the scrap of paper, but understanding.

I shut down the engine at the gas pump, still thinking. Choose the student, and give him time.

My chief instructor caught me before I was out of the airplane.

"You're back! We searched a solid week, looking for you from here to Cheyenne! We thought you were dead!"

"Not dead. Not dead at all. Just coming alive," I said. And beginning a tradition, I added, "Sir."

South to Toronto

The reason that a lot of adventures begin in this world is that the adventurers sit by the fire in comfortable living rooms and they haven't the faintest mist of an idea of what they are letting themselves in for. They stretch out in that easy chair and there is no such thing as cold or wet or wind or storm and they say well it's about time somebody discovered the North Pole and they lapse into a dream of glories and an hour later, dreaming still, they set wheels turning, maps unfolding, cogging other warm adventurers' lives to changing, to saying "Why not?" and "Jove! It should be done! Count me in!"—themselves tranced in a fantasy where hardship and trouble are only words that faint hearts look up in dictionaries.

Poke the fire, then, sit here in this warm chair, and let me spin you an adventure.

BARNSTORM WINTER CANADA!

What a sight, all those little towns snowed north of America, huddling through a white-quartz winter waiting for somebody to drop down from the sky and bring them colors and thrills in ten-minute hops to see their town from the air, three dollars the ride! And what a sound—that soft virgin February sighing to the touch of our skis! None of the problems of summer barnstorming here, no endless searching for pastures and hayfields smooth enough and long enough and close enough to town . . . why, all the world will be a place to land! Lakes are frozen flat, bigger than a hundred Kennedy Airports; every field that's rough in summer, or planted in tender crops, is a smooth perfect runway for our Cubs. Let's prove there's still room in the world for man individual, man challenging Canada winter to do its worst to keep him from bringing the gift of flight into the lives of those who have never been off the ground! How about it? The Canadians, after all, are frontiersmen, up there, with red-checked mackinaws and blue wool caps; axe in one hand, canoe in other, laughing all the time at danger—no hesitating there to buy our tickets! We'll fly up there for February, be home by March with the wilderness a part of our soul, the frontier alive again within us, the way it used to be!

That was all I had to spin to myself to be convinced. That, and letters from Glenn Norman and Robin Lawless, Canadians, woodsmen turned airplane pilots, no doubt, inviting me to stop by Toronto, someday.

Toronto! What a sound! A real Canadian outpost in the snowfields, Utopia for barnstormers! I stirred from the fire and got out the maps.

Toronto looks a little larger than one expects a wilderness outpost to look, but beyond it there are thousands of much smaller wilderness outposts, for miles around. Fenelon Falls, Barrie, Orillia, Owen Sound, Pentanguinishe. There are a dozen towns on the shores of Lake Simcoe alone, thirty miles from Toronto, and they are mere doorways to the teeming villages north and east and west. Imagine waking in the dawn, looking out from your warm sleeping bag under the wing, and finding the sign there in the ice:

PENTANGUINISHE!

My reply to the Canadians went out by return mail . . . would they be interested in joining the Winter Wonderland Flying Circus as wilderness guides? The wheels of adventure had begun to turn.

I sent letters the same day to American pilots with light planes and skis, mentioning that space was available in Canada for February.

Russell Munson signed on, with his Super Cub, the moment he got the news. All at once we had a starting date; on January 29th our two planes would touch their skis in Toronto, on January 30th we'd be off north, into high adventure.

We prepared all through January. I found a pair of used Cub skis in a hangar on Long Island, Munson found a pair of new ones in a factory in Alaska. We went through the flight over and again in his New York office—what must we be sure to take along?

Warm clothes, of course, and before a week was by we were clomping around the airfield in parkas and multilayer wools and insulated snowboots. Wing and engine covers, and we were enveloped in yards of sheet plastic and burlap, sewing them together just so. Hand-warmers for us, engine-warmers for the Cubs, inflatable tents, space blankets, survival kits, maps, spare parts, tools, cans of oil, sleigh bells for the skis. It is remarkable how much equipment one needs for a simple Canadian wilderness barnstorming tour.

My airplane was painted in enamel milk, which would never do; what customer would notice a white Cub parked on a snowbank? For the next three days I laid masking tape in candy stripes across the top of the wings and tail while Ed Kalish sprayed bright red over it all and remembered his days mechanicking at God's Cape, north of Hudson Bay.

"Got there one day," he said from a scarlet cloud of Dulux, "and it was seventy degrees below zero!"

My parka, the warmest garment I owned, was rated to fifty below.

"Had to start the engines with blowtorches up the exhaust stacks, turning the props backwards and getting the cylinders warm through the valves."

I went out that day and bought a propane blowtorch. And figured if I had to, I could stuff my parka with leaves.

Of the two other pilots I had invited, one wrote to say that he felt that Canada in February might be a little chilly . . . hadn't I meant we'd barnstorm Nassau?

When I finally replied that this flying circus was heading north, he wished me luck. I remember thinking that there was a strange reason to cancel adventure, because it would be cold. He had advised me to recall that the Cub had no cabin heater at all, but somehow that bounced off me like moose off ice.

The other pilot, Ken Smith, would meet us in Toronto on January 29th.

That gave us three Cubs, three pilots, and a pair of wilderness guides. We needed one more airplane, a Canadian, to join us so that we could be a true international circus, but I had no doubt that there would be dozens of CF aircraft ready to go along when we arrived in their country.

By mid-January the lakes were turning to ice all over Canada. New England ski resorts had opened for business and a few large snowflakes fell on Long Island.

On the night of the 20th I practiced sleeping among those flakes. It was only twenty degrees Fahrenheit outside, quite a bit warmer than we'd be having in Canada, but any test was better than none. Twenty degrees, I discovered, is actually quite cold. This was discovered some time around three in the morning. It wasn't that the tent had failed, or the space blanket wasn't working, but that the cold, after waiting that long, comes around and attacks the sleeper through the ground. I could think warm, all right, and fight it, but it took such a concentrated effort conjuring Saharas and bonfires that there was no time left for sleeping. At four I gave up and dragged tent and all back into the house. It was then that I began to think that while this was a lark for us, chasing this adventure, it was no game for winter. We were pointed dead-on into what the Air Force used to call a "survival situation" . . . men froze to death in warmer climates than February Canada! I packed an extra blanket at once.

Norman and Lawless flew to check out Lake Simcoe. The lake was frozen solid, the day they did, and the temperature was thirty below.

On January 27th, Toronto had its worst blizzard of the century. Towns were buried under snow, rescue operations were underway.

We were glad for this news; the deeper the snow, the closer we could land to town. When you are barnstorming, you might as well go home if you can't land close to town.

Extremely early on the morning of the 29th, Munson and I started engines under the dim place in the night that would be dawn . . . our engine exhausts were blue in the terrible stillness. It is around sunrise that adventurers reach that point at last where they begin to understand that they are out of their minds, just like everybody says.

"Russ, do you realize that this whole trip is folly? Do you realize what we are getting ourselves into? Look, I'm sorry I brought this whole idea up . . ." I wanted to say it, but didn't have the courage. Adventurers are cowardly about things like that.

Munson wasn't saying anything either, as the sky lightened and our engines warmed, and at last we climbed without words into our airplanes, taxied over the deserted concrete, and took off north, across Long Island Sound, across Connecticut. The outside air temperature at five thousand feet was eighteen degrees below zero, though I must admit that in the unheated cockpit it didn't feel any colder than ten or fifteen below. In the first place, I couldn't believe that I was going to spend a *month* in that temperature; in the second, I was thinking about summer, when the roads get so hot you can't walk without shoes and butter turns into yellow pools if you leave it out.

At our first stop, our very first stop, I noticed that my engine seemed to be blowing a little oil out the breather pipe. It always lost some oil, but this was more than usual. I unhooked the extension and let the pipe breathe in the warm engine compartment.

Since his airplane had a gyro compass, VOR and ADF radios, Munson was the flight leader to Toronto. My one magnetic compass was as sensitive to direction as a bench anvil, so I merely flew along as wingman and enjoyed the scenery, which was white and soft. Why this strange feeling, then, an hour after our second takeoff, that this was not the way to Canada at all? Those mountains to the right, weren't those the Catskills? And shouldn't the Hudson River be to our left? I moved into close formation and pointed to my map, looking a question at the flight leader. He looked at me and raised his eyebrows.

"Russ!" I shouted, "Aren't we heading south? *We're heading SOUTH!*" He couldn't understand what I was yelling, so at last I fell back and followed uncomplaining, as a wingman should, to see where he was going. He's been flying for ten years, I thought, so it must be me that's wrong. We're just following a different river. I noticed that he was checking his map, and this was reassuring to me. He didn't change course. We must be headed north . . . it's me that's lost, not for the first time.

But after a while it began getting warmer. There was less snow, down there on the ground.

The Super Cub realized, with a jolt, that somehow there had been a terrible mistake. It banked sharply right, changed course one hundred sixty degrees, and then drifted down to land at a little airport by the river. It was the Hudson, all right. For once in my life I was lost and it wasn't my fault!

"You may live this down," I told him gently, when we

had landed, "but believe me, it is going to take a long time . . ."

I was sorry at once, for he was deeply upset.

"I don't know what's the matter with me! I was following the highway and I noticed that the compass was a little off and the VOR wasn't quite right, but I was sure it was the highway! I just sat there and didn't pay any attention. I saw the compass, but I didn't pay any attention!"

It was not hard to change the subject. There was oil all over the belly of my airplane, blown out in the last hour. The landing gear and cowling were covered with it, congealed and frozen everywhere. A broken ring, perhaps, a cracked piston? We talked about turning back to check it over, but it sounded like the quitter's way.

"Let's press on," I said. "It's probably just suction there at the end of the breather pipe, taking out more than it should."

Munson nailed the course north on the Hudson, turned left at Albany, drove dead-on toward Toronto. An hour past Albany my oil pressure dropped one psi, then two. I have never had the oil pressure drop in any airplane engine without something bad happening soon . . . I pointed "down" to my flight leader and we landed at the next airport, five minutes away.

Another quart gone. The prospect of forty hours flying over the Canadian wilderness with an engine spraying its lifeblood into the sky was not the adventure I had chosen to play. It is one thing to be ready for engine failure, barnstorming, but quite another, and not so wise, I thought, to be convinced of it. Proceed or return, I was going to be a quitter; but better to be a warm quitter than a cold one perched in some Pentanguinishe treetop. Besides, the weather people told us, there's a fresh blizzard at the border.

I filled with oil and took off south, puzzled that I should be sad at missing out on a freezing. Once one gets started on an adventure, no matter how crazy a thing it is, the only way to rest easy is to carry it through, no matter what.

An hour and a half later the oil pressure fell five pounds, then ten, and then clunked against the peg at zero, leaving me to glide down the runway we had started from before dawn.

The problem with the engine was not as simple as a

cracked piston or a broken ring. The problem was that the cylinders had all worn oversize, beyond tolerance even for chrome plating. Four overhauled cylinders were available, at eighty-five dollars each, plus rings at thirty-two dollars and gaskets . . .

By the time I collected the money for engine parts, spring had come to Canada. Snow melted to grass, fields to crops, lakes dissolved from ice into blue water.

How's *that* for adventure? The winter raging wild up in Canada and you can challenge it and call it names and still sit all month by the fire . . . here's to adventure and adventuring! And next year, by Ned, it's on to the Pole!

Cat

It was a cat, a gray Persian cat. It had no name and it sat very carefully in the tall grass at the end of the runway, studying the fighter planes as they touched down in France for the first time.

The cat did not flinch as the ten-ton jet fighters whistled airily by, nosewheels still in the air and drag chutes waiting to spring from their little houses beneath the tailpipes. Its yellow eyes watched calmly, appraising the quality of the touchdowns, angled ears listening for the faint pouf! of the late-blossoming drag chutes, head turning serenely after one landing to watch the final approach and touchdown of the next. Now and then a touchdown was hard, and the eyes narrowed ever so slightly for an instant as the soft paw-pads felt the jar of airplane and soil for an instant as an airplane did not correct for crosswind, and great gouts of blue rubber-smoke angled from tortured wheels.

The cat watched the landings for three hours in the cold of an October afternoon, until twenty-seven airplanes had landed and the sky was empty and the last whine of dying engines had faded from the parking revetments across the field. Then the Persian stood suddenly, and without even a feline stretch of graceful body, trotted away to disappear in the tall grass. The 167th Tactical Fighter Squadron had arrived in Europe.

When a fighter squadron is reactivated after fifteen years of nonexistence, there are a few problems. With the barest nucleus of experienced pilots in a squadron of thirty, the 167th's problems centered around pilot proficiency. Twenty-four of its air crews had been graduated from gunnery training schools within the year before reactivation.

"We can do it, Bob, and do a good job of it," said Major Carl Langley to his squadron commander. "This isn't the first time I've been an operations officer, and I tell you I've never seen a bunch of pilots who are more eager to learn this business than the ones we have right here."

Major Robert Rider pounded his fist lightly against the rough wooden wall of his office-to-be. "That point I will grant you," he said. "But you and I have a job cut out for us. This is Europe, and you know European weather in the winter. Aside from our flight commanders, young Henderson has more weather time than any other pilot in the squadron, and he only has eleven hours of it. Eleven hours! Carl, are you looking forward to leading a four-ship flight of these pilots, in old F-84s, through twenty thousand feet of weather? Or to a GCA touchdown on a wet runway in a crosswind?" He glanced out the dirt-streaked window. High overcast, good visibility beneath, he noted unconsciously. "I'm going to run this squadron, and I'm going to run it well; but I'll tell you that I can't help but think that before the new 167th is a real combat-ready outfit, a couple of our boys are going to be scattered across the sides of mountains. I'm not looking forward to that."

Carl Langley's ice-blue eyes sparkled with the challenge. He was at his best doing the job that anyone else would have called impossible. "They've got the knowledge. They probably know instrument flying better than you and I, they're so fresh out of school. All they need is experience. We've got a Link. We can run that thing ten hours a day and fill our pilots with every instrument approach for every base in France. They volunteered to join the 167th, and they want to work for the squadron. It's up to you and me to work 'em."

The squadron commander smiled suddenly. "When you talk like that, I can almost accuse you of being eager yourself." He paused, and then spoke slowly. "I remember the old 167th, in England in 1944. We had the new Thunderbolt then, and we painted our little Persian battle-cat on its side. We weren't afraid of anything the Luftwaffe could get into the air. Eager in peace is brave in war, I guess." He nodded to his operations officer. "Can't say that I think we won't have our share of in-flight emergencies with this old airplane or that we won't need a lot of good luck before the boys start giving a meaning back to this squadron," he said. "But draw up your Link and flying schedules starting tomorrow and we'll begin to see just how good our youngsters really are."

In a moment Major Robert Rider stood alone in his darkening office, and he thought of the old 167th. Sadly. Of Lieutenant John Buckner, trapped in a burning Thunderbolt, who still attacked a pair of unwary Focke-Wulfs and took one of them with him into the hard ground of France. Of Lieutenant Jack Bennett, with six kills and glory assured, who deliberately rammed an ME-109 that was closing to destroy a crippled B-17 over Strasbourg. Of Lieutenant Alan Spencer, who brought back a Thunderbolt so badly damaged by cannonfire that he had to be freed from the wreckage of his crash landing by a crew with cutting torches. Rider had seen him after the crash. "It was the same '190 that got Jim Park," he had said from the whiteness of the hospital bed. "Black snakes down the side of the fuselage. And I said, 'Today, Al, it's going to be you or him, but one of us isn't going to make it home.' I was the lucky one." Alan Spencer volunteered to go back into combat when he was released from the hospital, and he did not return from his next mission over France. No one heard him

call, no one saw his airplane hit. He simply didn't come back. Despite their battle-cat insignia, the 167th pilots did not have nine lives. Or even two.

Eager in peace is brave in war, Rider thought, looking absently at the scar along the back of his left hand, his throttle hand. It was wide and white, the kind of scar left only after an encounter with a Messerschmitt's thirty-caliber machine-gun bullet. But eagerness is not enough. If we're going to make it through the winter without losing a pilot, we'll need more than eagerness. We've got to have skill and we've got to have experience. So thinking, he walked outside into the overcast night.

The days whipped quickly by for Second Lieutenant Jonathan Heinz. All this talk of weather and look out for Europe in the winter was nonsense, sheer nonsense. November was bright and spilling with sun. December was ready to spring onto the calendar and the base had had only four days of low ceilings, which the pilots spent working on the ops officer's latest instrument quiz. Major Langley's instrument quizzes had become a standard of the squadron; a new one every third day, twenty questions, one wrong answer allowed. Fail a test and you stay another three hours at the flight line with the instrument manuals, until you pass the alternate test, one wrong answer allowed.

Heinz pressed the starter switch of his aging Thunderstreak, winced in the concussion of a good start, and taxied to the runway behind Bob Henderson's airplane. But that's the way to get to know instruments, he thought. At first everyone was staying the three hours and cursing the day that they volunteered for the 167th Tactical Fighter Squadron. Tactical Instrument Squadron, they called it. Then you got the knack of it, and it seemed somehow that you knew more and more of the answers. It was pretty rare now to have to stay the three hours.

There was a little thud in the engine's roar when Heinz retracted his engine screens before takeoff, but all the engine instruments showed normal, and strange noises and little thuds are not unusual in the F-84. Oddly enough, though, at a time when he usually noticed little but the instruments and the leader's airplane rocking firmly against full throttle and locked brakes, Jonathan Heinz noticed a gray Persian cat sitting calmly at the edge of the runway a few hundred feet ahead of his airplane. Cat must be completely deaf, he thought. His engine, linked to the thick black throttle under his left glove, crackled and roared and spun blue fire through stainless-steel turbine blades to unchain seventy-eight hundred pounds of thrust within his airplane.

He was ready to roll, and he nodded to Henderson. Then, for no reason, he pressed the microphone button under his left thumb on the throttle. "There's a cat out on the edge of the runway," he said into the microphone set into his green-rubber oxygen mask. There was a short silence.

"Roj on the cat," Henderson said seriously, and Heinz felt foolish. He saw the mobile control officer in his miniature control tower at the right side of the runway reach for his binoculars. Why did I say a dumb stupid thing like that, he thought. I will not say one more word this flight. Radio discipline, Heinz, radio discipline! He released his brakes at the nod of Henderson's white helmet, and the two airplanes gathered a great reserve of speed and lifted into the air.

Eight minutes later Heinz was talking again. "Sahara Leader, I got an aft overheat light and the rpm's surging about five percent. Power's back, light's still on. Check me for smoke, will you?" What a calm voice you have, he thought. You talk too much, but at least you're calm. Sixty hours in the '84 and you should be calm. Take it easy now and try not to sound like a little kid on the radio. I'll turn around and drop the external tanks, fly a simulated flame-out pattern, and land. I couldn't be on fire.

"No sign of smoke, Sahara Two. How's it doing now?"

Calm voice, Heinz. "Still surging, Leader. Fuel flow and tailpipe temp are going back and forth with it. I'm going to drop the tanks and land."

"OK, Two, I'll keep an eye for smoke and handle the radio calls if you'd like. But be ready to jump out of the bird if she starts to burn."

"Roj." I'm ready to jump out, Heinz thought. Just raise the ejection seat armrest and squeeze the trigger. But I think I can get the airplane back down all right. He listened to Henderson declare an emergency, and as he descended slowly into the flameout pattern he saw the square red fire trucks burst from their garages and race to their alert slots at the taxiways. He could feel the engine surge in the throttle. It will be sort of touch-and-go here. I'll drop the tanks on final approach before I get down to five hundred feet, I'll pull the nose up and eject. Below five hundred feet, I'll have to take it on in, no matter what. He brought the throttle back to give an engine speed of fifty-eight percent rpm, and the heavy airplane dropped more quickly through the pattern. Flaps down. I have the field made, for sure . . . Gear down. The wheels locked in place. He passed through four hundred feet. Thud. Thudthud. A big surge.

"There's a lot of smoke from your tailpipe, Sahara."

Wouldn't you know it! This thing's going to explode on me, and I'm too low to bail out. What do I do now? He pressed the drop tank jettison button and the airplane bounced a little as four thousand pounds of fuel fell away. A harsh grinding from the engine, behind him. He noticed, suddenly, that the oil pressure was at zero.

A frozen engine, Heinz! You got no flight control with a frozen engine. What now, what now? The control stick went solid and immovable in his gloves.

The officer in mobile control did not know about the frozen engine. He did not know that Sahara Two would make a gentle roll to the right and strike the ground inverted, or that Jonathan Heinz was helpless and committed to die. "You have a cat by the runway," the mobile officer said, with the mild relaxed humor of one who knows that danger has passed.

And it came to Heinz suddenly. In a burst of light.

Emergency hydraulic pump, the electric pump! His airplane was beginning to roll, a hundred feet in the air. His glove smashed the pump switch to EMERG, and the stick came alive again, quickly. Wings level, nose up, nose up, and a beautiful touchdown in front of mobile. At least it felt beautiful. Throttle off, drag chute out, fuel off, battery off, canopy open and be ready to jump out of the thing. The giant square fire trucks, scarlet lights blazing atop their cabs, roared along next to him as he slowed through thirty knots on the landing roll. His airplane was completely quiet, and Heinz could hear the truck engines, sounding like great inboard cruiser engines, laboring in high gear. In a moment he had rolled to a stop, unstrapped from the cockpit, and jumped down to stand behind a fire truck that hosed thick white foam on a broad patch of discolored aluminum aft of the wing root.

The airplane looked forlorn, unwilling to be the center of such concentrated attention. But it was down, and it was in one piece. Jonathan Heinz was very much alive, and not a little bit famous. "Nice going, ace," the pilots would say, and they'd ask him about how it felt and what he thought and what he did and when, and there would be the routine accident investigation and there could be no other conclusion than well done, Lieutenant Heinz. No one would guess that he came within a few seconds of dying because he had completely forgotten, like a brand-new pilot, about the emergency hydraulic pump. Completely forgotten . . . and what had reminded him? What had snapped his thought to the red-covered switch at the last instant that it could save him? Nothing. It had just come to him.

Heinz thought some more. It had not just come. Mobile control told me about the cat by the runway, and I remembered the pump. There's an odd one for you. I'd like to meet that cat.

He looked down the long white runway. He could see no cat. Even the mobile control officer, with his binoculars, could not then have seen any cat. The squadron was later to ride him unmercifully about his unlucky cat, but at that moment, by the runway or across the whole of the base, there was no such thing as a gray Persian cat.

It happened again, less than a week later, to another second lieutenant. Jack Willis had almost finished his first simulated combat mission after completing his checkout in the

F-84. It had been a good mission, but now, in the landing pattern, he was worried. Twenty-knot crosswind, where did that come from? It was ten knots down the runway when we took off, and now it's twenty knots across. He rolled his airplane level on the downwind leg of the pattern. "Say again the wind, please, tower," he called.

"Roj," the tower's last explanation was entirely unnecessary. The wind was as cross as it could possibly be.

"OK, Two, let's watch the crosswind," said Major Langley, and called, "Eagle Lead is turning base, gear down, pressure up, brakes checked."

"Cleared to land," the tower operator replied.

Willis reached forward with his left glove and slammed the landing gear lever to DOWN. OK, OK, he thought, this will be no problem. I'll just keep the right wing way down through the round-about, touch on the right wheel, and follow through with plenty of rudder. Plenty of rudder.

He turned toward the runway, and pressed the microphone button. Haven't run off a runway yet, and I don't intend to do it today. "Eagle Two is turning base . . ." The right main gear indicator, the green light that should have been shining, was out. Left main was locked down, nosegear was locked down. But the right main gear was still retracted. The red warning light in the transparent plastic landing gear handle was shining, and the squeal of the gear unsafe warning horn filled the cockpit. He heard the horn in his own earphones as he held the microphone button down. On their radios, the tower operators would hear the horn. He lifted his thumb, then pressed it down again. "Eagle Two is going to make a low approach; requests a gear check by mobile control."

An odd feeling, to have something wrong with the airplane. The landing gear usually works so well. He leveled at one hundred feet over the runway and flew past the miniature glass tower. The mobile control officer stood outside, in the blowing waves of autumn grass. Willis watched him for a second as he passed. The mobile control officer was not using binoculars. Then he was gone, and the solitary F-84 whipped over the far end of the runway, above Eagle Lead, safely on the ground.

"Your right main gear is up and locked," the voice came flatly from mobile control.

"Roj. I'll cycle the gear." Willis was pleased with his

voice. He climbed slowly to one thousand feet, raised the landing gear, and lowered it again. The right main "safe" light remained stubbornly out, and the warning light in the plastic handle persisted redly. Another fifteen minutes of fuel. Four times Willis recycled the landing gear, and four times the right main gear indicated unsafe. He pulled the handle out a half inch and pressed it to EMERG DOWN. There was a faint click from the right, but the condition was the same. He was concerned. There was no time for the fire trucks to lay a strip of foam down the runway, if he was forced to land with the right gear still locked up. To land with it up on a hard dry runway, crosswind, would be inviting an end-over-end cartwheeling crash as soon as the unwheeled wing touched the concrete. The only alternative was bailout. Here's a decision for you, he thought. And irrationally: one more flyby, maybe the gear is down now.

"It's still up," the mobile control officer said, before Willis had even flown by the miniature tower. The grass was waving greenly, briskly, and he noticed suddenly at the edge of the runway a small gray dot. With a shock of surprise, he realized that it was a cat. Heinz's lucky cat, he thought, and for no reason he smiled under his oxygen mask. He felt better. And a thought came from nowhere.

"Tower, Eagle Two is declaring an emergency. I'm going to make one more pass around; try to bounce on the left gear to knock the right one down."

"Understand you are declaring an emergency," the tower replied. The tower was primarily concerned with meeting its responsibility, which was to ring the bell that sent crash crews scrambling for the red trucks. Responsibility met, the tower became only an interested observer, and very little help.

Jack Willis, oddly, felt like a new person, enormously confident. The bounce on a left wheel in a strong right crosswind was a trick of coordination reserved for thousand-hour pilots, and Willis had just over four hundred hours in the air, sixty-eight in the F-84.

Those who watched the next approach called it the work of an old-time professional pilot. With left wing down, with hard right rudder, with controls only moderately responsive at landing airspeed, Second Lieutenant Jack Willis bounced his twenty-thousand-pound airplane six times on its left main landing gear. On the sixth bounce, the right gear swung suddenly down and locked into place. The third green light came on.

The crosswind landing that followed was simple by comparison, and his airplane touched smoothly down on its right wheel, then its left, and last of all, the nosewheel. Full left rudder in the landing roll and a touch of left brake as the airplane slowed and tried to weathervane into the wind, and the emergency was over. The crash crews in their bulky white suits of asbestos were unnecessary and out of their element in the normalcy that followed. "Nice job, Eagle Two," mobile said simply. And the gray Persian cat, that had watched the landing with uncatlike, one might almost say, professional interest, was gone. The 167th Tactical Fighter Squadron was gradually pulling itself into fighting shape.

The winter came. Low clouds moved in from the sea to become a permanent companion of the hilltops that surrounded the airbase. It rained much, and as the winter wore on, the rain became freezing rain and then snow. The runway was icy and drag chutes and very careful braking were necessary to keep the heavy airplanes on the concrete. The tall emerald grass turned pallid and lifeless. But a fighter squadron does not cancel its mission each winter, there is always flying and training to be done. There were incidents as the new pilots were faced with unusual aircraft problems and low ceilings, but they had been trained well on instruments and somehow the gray Persian cat sat carefully at the edge of the runway as each of the afflicted airplanes landed. The Persian became known to the pilots simply as "Cat."

One freezing afternoon, just as Wally Jacobs touched down uneventfully after a hydraulic system failure and a no-flap, no-speedbrake approach through a five-hundred-foot ceiling, Captain Hendrick, on duty as mobile control officer, ventured to capture the cat. It sat quietly, looking down the runway, absorbed in watching Jacobs's airplane after it whistled past. Hendrick approached from behind and gently lifted the cat from the ground. At his first touch it became a ball of gray lightning. There was an instant slash of claw along Hendrick's cheek, and the Persian streaked to the ground and away, disappearing at once in the tall dry grass.

Five seconds later the brakes on Wally Jacobs's airplane failed completely, and he swerved off the runway at seventy knots into the not-quite-frozen dirt. The nosewheel strut sheared immediately. The airplane disappeared in a great sheet of flying mud, slewed to collapse the right main landing gear and split the droptank, and slid around, backward, for another two hundred feet. Jacobs left the cockpit at once, forgetting even to close the throttle. In a second, as Hendrick watched, the airplane burst into brilliant flame. It burned fiercely, and with the airplane was destroyed a record for flying safety unmatched by any other fighter squadron in Europe.

The findings of the investigation were that Lieutenant Jacobs was at fault for allowing the airplane to leave the runway and for neglecting to close the throttle, allowing the still-turning engine to ignite the fire. If he had not forgotten, like a grossly inexperienced pilot, to stopcock the throttle, the airplane would have been able to fly again.

The board's decision was not a popular one with the 167th Tactical Fighter Squadron, but the cause of the destruction of the airplane was laid to pilot error. Hendrick mentioned the cat, and an order, unwritten but official, was sent through the squadron: the Persian was never to be approached again. From that moment, Cat was rarely mentioned.

But once in a while, as a young lieutenant brought an ailing airplane down through the weather, he would ask of mobile control, "Cat there?" And the mobile control officer would scan the runway edge for the sculptured gray Persian, and he would pick up his microphone and say, "He's there." And the airplane would land.

Winter wore on. The young pilots became older, absorbed experience. And as the weeks went by, Cat was seen less and less frequently at the edge of the runway. Norm Thompson brought in an airplane with the windscreen and canopy completely iced over. Cat was not waiting by the runway, but Thompson's GCA was a professional one, born of training and experience. He made a blind touchdown, jettisoned the canopy to be able to see, and rolled to an uneventful stop. Jack Willis, now with one hundred thirty hours flying experience in the F-84, came back with an airplane heavily damaged by ricochets picked up after a firing run at a new strafing range laid over a base of solid rock. He landed smoothly, although Cat was nowhere to be seen.

The last time Cat appeared by the runway was in March.

It was Jacobs again. He called that his oil pressure was falling, and that he was trying to make it back to the field. The ceiling was high, three thousand feet, when he broke into the clear after a radar vector and called the runway in sight.

Major Robert Rider had raced his staff car to mobile control as the notice of emergency in progress reached him. This is it, he thought. I'm going to see Jacobs die. He closed the glass door behind him as the pilot asked, "Cat happen to be down there?"

Rider reached for the binoculars and scanned the edge of the runway. The Persian sat quietly waiting. "Cat's here," the squadron commander told the mobile control officer seriously, and seriously the information was relayed to Jacobs.

"Oil pressure zero," the pilot said matter-of-factly. Then, "Engine's frozen, the stick is locking. I'll try to make it on the emergency hydraulic pump." A moment later he said, suddenly, "No I won't. I'm getting out." He turned his airplane toward the heavy forest to the west and ejected. Two minutes later he was sprawling in the frozen mud of a plowed French field, his parachute settling like a tired white butterfly about him. It was over that quickly.

The investigation board was to find later that the airplane struck the ground with both hydraulic systems completely locked. The emergency hydraulic pump failed before impact, they discovered, and the airplane hit with controls frozen and immovable. Jacobs was later to be commended for his judgment in not attempting to land the stricken airplane.

But that was to be later. As Jacobs's parachute drifted down behind a low hill, Rider leveled the binoculars at the gray Persian, who stood suddenly and stretched luxuriously, claws digging into the frozen earth. Cat, he noticed, was not a perfect sculpture. Along his left side, from ribs to shoulder, ran a wide white scar that the battle-gray fur could not cover as he stretched. The graceful head turned as Rider watched, and the amber eyes gazed squarely at the commander of the 167th Tactical Fighter Squadron.

The cat blinked once, slowly, one might almost say amusedly, and walked to disappear for the last time in the tall grass.

Tower 0400

I closed the door behind me just as the twenty-four-hour clock by the light gun ticked through 0300. It was dark, of course, in the tower, but it was a much different sort of darkness than the black I had just stepped in from. That dark was a thing that anyone could use for any purpose; for cards or for crime, or for the war hinted and threatened in the headlines downstairs.

The darkness in this aerie of glass and steel was a specialized dark. Everything it touched had the same air of professional purpose about it; the clock, the lightly hissing radio receivers in their bank along one low wall, the silent never-ending sweep of the pale green line of the radar scope. It was a professional darkness to shroud the world of people who fly airplanes. There was no malice in this dark, it was not there to drag the airplanes down or to make it difficult for them to fly. It was just a matter-of-fact, businesslike, I-am-here-now darkness. The beacon rotating with its busy hum a few feet overhead was not turning to fight this dark, but to pinpoint a landing field on a map of black.

The two tower operators who worked the graveyard shift were expecting me, and extended hands from behind the orange glow of their cigarettes. "What brings you up here at this hour?" one asked quietly. All the talk on this shift was quiet, as if to keep from waking the city that slept at our backs.

"Always wondered what it was like," I said.

The other man laughed, again quietly. "Now you know," he said. "This one minute is a pretty good example of what it's like all through the shift."

The static hissed lightly on in the speakers, the light gun hung unswinging from the ceiling, and the pale line of the radar turned endlessly, tirelessly. The airport was waiting. At that moment, somewhere out in the starred night sky, an airliner bored steadily ahead, her long aluminum nose pointing at the field guarded by this tower. It wasn't even an image yet on the farsighted eye of the radar, but the first officer was calling for weather at our field and leafing through his briefcase for let-down plates. His engines roared steadily on in the darkness outside, and the needles of the oil quantity gages had dropped down, confirming the length of the flight.

But in the tower the air was quiet and still. The blue stars that were the taxiway lights stood frozen in their orderly constellation on the field, waiting to guide any pilot who thought of taxiing at this hour.

Down on the lightplane ramp, a flashlight snapped on, making a little yellow eye on the concrete with its short beam. As I watched, the eye jumped up the trim fuselage of a Bonanza, found the door, and disappeared into the cock-pit. It reappeared in a moment, and for a second I saw the shadowy form of the pilot who held the light as he stepped off the wingwalk.

The tower operators continued their quiet talk together about the places they had been and the things they had seen. I watched the eye of the flashlight, fascinated. Where was that pilot going? Why did he leave so long before the sun? Is he a transient pilot going home or a home-town pi-lot leaving?

The little pool of yellow light stayed for a moment on the aileron hinges, splashed down the leading edge of the right wing, disappeared beneath it into the wheel wells. It ap-peared suddenly on the cowling and waited patiently until the dzus fasteners were half-turned open and the cowl lift-ed. It jumped eagerly onto the engine, checking the spark-plug terminals, the oil level; it wandered for a moment where it pleased around the finned cylinders and the low-swing engine mount. The cowl came back down and fas-tened shut. The light brightened as it moved the length of the propeller, and was gone for a minute on the other side of the airplane. It appeared again on the fuselage and slipped into the cockpit.

The flight line was as dark as it had been when I came, but out in its dark now was a man, and he was getting his

airplane ready to fly. In the binoculars I found the faint glow of the dim cockpit lights as they came on, and in a minute the red-green of his navigation lights flashed, giving dimension to his machine. And suddenly the silence in the aerie was broken.

"Tower, Bonanza four seven three five Bravo on the ad ramp, taxi for takeoff." The voice stopped as suddenly and abruptly as it had begun.

In our high glass cube the smooth professional voice of the tower operator answered, speaking into his microphone as if this were the thousandth call he had answered this morning, instead of the first.

A single brilliant white light slammed into the darkness of the ramp below, and in its white the concrete showed its own true white and the yellow of its painted line. The bright light moved easily through the blue constellations of the field, homing on the end of the runway's long strip of white lights. It stopped, and flicked out. Even in the binoculars the cockpit lights were too dim to see; the only evidence of the plane's form was a short break in the orderly row of blue taxiway lights.

In a minute our quiet air was broken again by the voice from the VHF speaker. "Tower, three five Bravo; think you can work me in for a takeoff?"

"Wise guy," the controller said, reaching for his microphone. "Might be able to squeeze you onto the schedule, three five Bravo. Cleared for takeoff, wind calm, no reported traffic."

"Roj, tower, three five Bravo's rolling."

The black blot against the lights moved ahead as he talked, the only motion on a still field. In fifteen seconds the field lights shone as before, and a flashing green navigation light reached for the dark horizon.

"Beautiful night," the pilot said thoughtfully to the VHF. The field was still again.

Those were the last words we heard from three five Bravo, and his lights faded into the night. I never will know what field he calls home, or where he was going that night, or who he is. But in that one last call, still captured on the impersonal tape recording of the tower, the pilot of that Bonanza suggested that perhaps pilots really are different from all other men.

They share the same nontransferable experience of flying

alone; if they are also moved by the same beauty of the sky, they have too much in common to ever be enemies. They have too much in common to ever be less than brothers.

The field waited again, patiently, for the next airplane.

What a fraternity that would be, a real brotherhood of all the men who lift airplanes into the sky!

"This will be a Lufthansa flight coming in," the controller said, pointing to the disc on the radar scope.

Lufthansa was a blurred ellipse, a quarter of an inch wide, moving slowly in from the edge of the screen. He left a ghostly green luminous trail that made him look like a tiny comet pointed toward our tower at the center of the scope.

As we looked out from the tower's glass into the crystal night air, not a light moved in the sky. The comet closed on the center of the screen, the minute hand of the many-numbered clock swung around, and still the lights in the sky were stars.

Then all at once Lufthansa was a flashing red anticollision light in the distance, and her first officer pressed the mike button on his control wheel. "Tower, Lufthansa Delta Charlie Charlie Hotel, fifteen miles east for landing." The first officer spoke easily and precisely, and Lufthansa was pronounced "Looftahnza."

The thought swept me again. He could just as well have said, "Deutsche Lufthansa für Landung, fünfzehn Meilen zum Osten," and he would have been as much, or even a little more a brother of the fraternity than I, standing in the high tower.

What if every pilot knew, I thought, that we are already brothers? What if Vladimir Telyanin climbing the kicksteps of his MIG-21 knew it as well as Douglas Kenton in his Meteor and Erhart Menzel in his iron-crossed Starfighter and Ro Kum Nu tightening the shoulder harness in his YAK-23?

Lufthansa swung easily down the ILS glidepath, his landing lights shining like bright eyes looking for the runway.

What if the fraternity refused to fight among itself?

Lufthansa taxied close to the terminal building, and in the tower we listened to her four engines whine down into the quiet.

The radios hissed softly on, the sky was still again, the

green line of the radar scope agreed that we were alone once more in the darkness. As the hands of the clock by the light gun touched 0400, I said my thanks and goodbyes to the controllers and stepped to the iron grating and stairway outside. I felt that difference of the blackness again, the same dark that touched the pages of the newspapers down at the end of the stairway.

Above me, and above the field of sleeping airplanes, less one American lightplane, plus one German airliner, the long beam of the beacon swept around. Brothers. My leather soles rang on the metal stairs. At night, in the dark, you think funny things.

What if they all knew? I thought.

The snowflake and the dinosaur

Have you ever wondered how a dinosaur felt, trapped in a Mesozoic tarpit? I'll tell you how he felt. He felt exactly the way you would feel if you had force-landed in a winter hayfield in northern Kansas, fixed your engine, and tried to take off again through a carpet of wet snow. Helpless.

They must have tried and tried, those poor stegosaurs and brontosaurs, turning up full power, thrashing like crazy, sending tar flying all directions till sundown caught them in darkness and at last they got so tired it was a blessing to give up and die. It's the same way in snow, for an airplane—in a mere six inches of picturesque, level snow.

With sundown coming on and a long walk from nowhere, the pilot's alternative to dying is a cold night in a sleeping bag under shadow of new storms coming. Yet in spite of that, to me, the trap of snow wasn't fair. I didn't have time for it. Twenty tries at takeoff had won me only the understanding of the power of a snowflake, multiplied a thousand billion times. The heavy wet stuff turned to thick soup blurring underwheel, blasting violent hard fountains against the struts and wings of my borrowed Luscombe. Full power would drag us up to thirty-nine miles per hour at the fastest, and we needed forty-five minimum to take off. An atom-age dinosaur, we were caught in the wilderness.

Between tries, while the engine cooled, I walked the field, frowning at the injustice of it all, stamping down a narrow white runway, wondering if I'd be camping in the cockpit till spring.

Every new try at takeoff smashed the snow easily enough

under the wheels, but at the same time built walls alongside, in ruts a foot deep. Jerking in and out of those tracks was like trying to take off with a balky rocket engine bolted to the plane. In the rut, we accelerated like a shot, but swerve two inches and bam! the nose pitched down, I was thrown forward in the seat, and we lost ten miles per hour in a split second. It was a kind of fixation. Bit by bit, I thought, I've got to wear down a runway till we can finally take off; or else it's the rest of the winter here. But it was hopeless. If I had been a dinosaur, I would have laid me down to die.

When you fly old-time airplanes, you expect to have forced landings now and then. It's nothing special, it's part of the game, and no wise pilot flies an antique out of gliding distance of a place to land. In my few years flying, I'd had seventeen forced landings, not one of which I had even thought unfair, for all of which I was more or less prepared.

But this was different. The Luscombe I flew now was hardly an antique; it had higher performance than ultramodern planes of greater horsepower, and had one of the world's most reliable engines. I flew this time not for fun or for learning, but for a business trip from Nebraska to Los Angeles and return, and I was almost finished with the flight and this was no time for a forced landing. It was more a bother because the engine had never quit. The problem had been a fifty-cent throttle-linkage connection, snapped in two. So when the engine power fell back to idle rpm on the last leg of my business trip—with an appointment waiting in Lincoln—it was the first unfair forced landing I had ever had.

Now, having repaired the linkage, I couldn't get off the ground again, and it was just an hour till sunset, when dinosaurs must die.

For the first time in my life, I understood the modern-airplane pilots who use airplanes as business tools and don't want to be bothered with such things as aerobatic training and forced-landing practice. Chances are rare that they'll ever stop or that a minor little linkage will break in half. It is fair for that sort of thing to happen to a sport pilot, who pays attention to such esoteric trivia and enjoys being ready for it, but not for me in my business plane when I have people waiting for me at the terminal and a dinner planned for six p.m. sharp. Because a forced landing for a businessman is quite honestly unfair, I began to realize that he gets to thinking it can't possibly happen.

I planned to make one more try to get out of that little field in Kansas before dark. I was already late for my meeting, but the snow didn't care at all. Nor did the cold, or the field, or the sky. The tarpits hadn't cared about the dinosaurs, either. Tarpits are tarpits and snow is snow; it's the dinosaur's job to get himself free.

The twenty-first try at takeoff, then, the Luscombe, spraying snow, tracking down a rut just long enough, bounced to forty-five, shuddered, wallowed, staggered into the air, touched snow again, shook it off, and at last flew.

I thought about it all as we turned for Lincoln, scudding along over the shadows of dusk. I now had eighteen forced landings in my logbook, and only one of them was unfair.

Not a bad record.

MMRRrrrowCHKkrelchkAUM . . . and the party at LaGuardia

Do you come suddenly awake, ever, to find yourself standing on the rail of a monster bridge, or the ledge of some hundred-story office building, find yourself swaying, teetering out over empty space, and wonder how it's happened that you're there, ready to jump? And in answer, do you get a pushing volley of reasons all poking at you—wars here and hatreds there and dog-eat-dog across the way and the only thing that matters is the lousy buck and the meadows are all junkyards and the rivers are all slag and nobody cares about right instead of wrong and good instead of evil and gentleness instead of wrath, and chances are a mistake was made somewhere and you were in fact born into the wrong world, that this isn't the earth you applied for at all and the only way to change it is to hop off some high place with the wish that the ground below will be a doorway into other lives, and better ones, with challenge and joy and the chance for getting something worthwhile done?

Well, wait a minute before you jump. Because I have a story to tell you. The story is about a couple who are as crazy as sane folk in Bedlam, who just might be friends of yours. Who decided that instead of jumping they would grab the world and whack it a couple of times and make it turn out the way they want it to turn out.

The man is James Kramer, pilot. The woman is Eleanor Friede, an editor in a book publishing company. What they did to the world was to start an airline.

East Island Airways was founded because Jim Kramer saw a 1941 twin Cessna T-50 Bamboo Bomber going to ruin on an airport tie-down and he wanted to rescue it, he wanted to save it.

East Island Airways was founded because Eleanor Friede wanted a way from New York City to her Long Island beach house that would not strangle her dead in four hours of bumper-locked summer automobiles.

East Island Airways was founded because Mrs. Friede met Mr. Kramer when she learned to fly, and not long after he came running and shouting into her house that he had found a Bomber that had to be saved and he'd put up half the money if she'd put up half and they could do something with it to make it pay its way but just come out now, turn off the stove and come out now and look at this airplane and Eleanor if you don't think this is the most beautiful thing and maybe let's not think we'd make a lot of money but there must be other people who hate the traffic too and they could at least be enough, on the tickets, to break even and we could save the Bomber!

So it was that Eleanor Friede saw the old round-engine twin waiting there in the sunlight and she thought it was lovely and she liked it as much as Jim Kramer did, for its majesty and its charm and its style. It had all these things, and it cost seven thousand dollars, at a time when other Bombers sold for four thousand, and five. But other Bombers didn't need to be rescued from owners who did not love them and seven thousand dollars split is only thirty-five hundred dollars apiece. Then and there, East Island Airways was born.

There were already air taxi lines flying from New York's LaGuardia Airport to East Hampton, Long Island. So what.

The other air taxis had modern airplanes; they had several modern airplanes each. Imagine that.

The Bomber would have to be completely inspected and most likely rebuilt, and that would be expensive, that could take most of the money the two had saved all life long. Interesting.

There would be papers required, and work to form the company, to qualify for operating certificates, to calculate and buy insurance. Quite so.

Statistics show, logic shows, common sense shows without the smallest flicker of doubt that there would hardly be a dime's profit and more likely a dollar's loss, perhaps a many-dollars' loss. Remarkable.

Mr. Kramer was president and chief pilot.

Mrs. Friede was chairman of the board and secretary-treasurer.

Now this world that we live in, the world that occasionally drives us to our jumping-off places, did not particularly like this event. It did not particularly dislike it, either, but acted in the cold uncaring way that the world usually has, and began to put the screws to East Island Airways, with a certain blind curiosity, to see when it would crack.

"The cost of the airplane was the least of the expense," said Mrs. Friede, "absolutely nothing. I'll show you the books if you want to look at the books. I hid them."

Kramer worked five months over the airplane with a Long Island overhaul company, recovering the fuselage, installing radios, ripping out the old interior and pleating in a new one.

"You know the expression 'Never throw good money after bad'?" he said. "Well, we had one like it: '*Always* throw good money after bad.' We had planned to spend some money getting the Bomber in shape, but when we got the bill, it said *nine thousand dollars!* Nine thousand three hundred dollars. It was unbelievable. We sometimes sat at a table in a stupor, wondering . . . you know . . . hm." His voice trailed off, thinking about that, and the chairman of the board went on.

"Everybody, *everybody* warned us that we didn't have enough capital, and one airplane was a disaster for an airline, and it couldn't work. And they could prove it—they didn't have to prove it, we knew this. But neither of us was making our living from it, that was one thing. And if we were putting in money that we needed to pay bills or something . . . ah . . . well, actually we *were* putting in money that we needed to pay bills . . . but the bills waited and we didn't starve, somehow."

When the Bomber was at last ready to fly, *EIA* lettered calmly on its rudder, it had cost the partners sixteen thousand five hundred dollars. Split, was only eight thousand two hundred fifty dollars each. But the money wasn't lost, the savings hadn't disappeared. East Island Airways had an airplane!

A Parlor Plane service to the Hamptons—
for not too many people.

You are invited to be a charter member of

EAST ISLAND AIRWAYS

> *East Island Airways is one beautiful, big, leather-lined twin Cessna. Not new. Not even very sleek (see photo). But fully FAA approved and a pampered beauty. Comfortable. The kind of no-climbing-over-passengers spaciousness that makes you think of a well-kept Packard limousine with all those miles of carpet. We depart from LaGuardia and cruise 140 mph to East Hampton in 45 minutes . . .*

The membership fee was one hundred dollars, and the fare was fifteen dollars each way, a hundred-mile flight.

It didn't work. Nobody joined. The world laid on its pressure curiously, listening for breaking noises.

"A lot of friends of Eleanor's expected to ride on the airplane for nothing, I'm sure. I think when people get an ad and you're flying, they think the outfit has a lot of money and what is one more person, more or less? In the beginning we didn't mind, we just sort of wanted to let them know we existed."

That was no breaking noise, and it sounded peculiar to a dog-eat-dog competitive world. Not many airlines fly passengers for nothing, just to let them know they exist.

"Business was very slow till the fourth of July, and then suddenly we started carrying a lot of people. We did everything by charter, people would call and charter the airplane. This actually worked out pretty well because we made enough friends at the beginning for that to be a busy three or four days out of the week. And there were charters to New England and Maine and so forth. We kept pretty busy."

Odd. The steel-eyed no-nonsense practical world laid on the pressure, and the only response sounded strangely like the noise of the world, cracking a bit.

"People were always expecting it to fall down and they wanted it not to work. It's old and it can't be, but it does and it keeps going and they don't know what to think, after a while. They don't know. They wonder if maybe things that are old are better than things that are new.

"A wooden airplane does not fatigue. They'll have problems with twin Beeches, they'll have problems with 310s, they'll all be in the scrap heap and they'll be there because of metal problems, when twenty years from now the guy will say, 'It's going to cost you a hundred thousand dollars to fix that metal airplane of yours,' and there'll be the Bomber sitting next to it, kind of chuckling to itself and saying, 'Don't you wish *you* had wooden spars?'

"We were able to make enough. People would say, 'Gee, it's great, you must be making a lot of money,' and I'd say, 'Sure, sure,' because I couldn't go into the thing that we were in fact not making a lot of money, people wouldn't understand.

"It was the kind of thing where you were beating the system. Everybody flying was trying to allow the passenger those fast airplanes, planes that had tremendous capability, and all the passengers got was crammed and smashed in and baggage on their noses and everything. Nobody else would think of running an airplane that old, and nobody thought it would last more than a week.

"They knew it at LaGuardia, after a while. In the beginning they couldn't figure out what it was—it was always, 'Say again your type aircraft?' If we were making an IFR approach coming down the localizer at ninety knots they'd say, 'What's a twin Cessna doing so slow? You can fly faster than that!' I'd say, 'Well, I could, but I couldn't put the wheels down.' They couldn't figure out this was an old *old* twin Cessna, not . . . they figured it was an old Cessna 310. 'No, it's an old old *old* twin Cessna,' and they'd say, 'Oh. OH! You mean *them!*' "

"Do you remember, Jimmy," asked the chairman of the

board, "we were landing and the tower said, 'Twin Cessna on final, is that a metal-wing aircraft?' And you said, 'Negative. Fabric wings.' And the guy said, 'Gee! They sure do shine!' "

"Yeah. We'd be talking to a controller and he'd say, 'Hey, I had an uncle who flew them during the war,' and he'd say, 'Boy . . .' and at this time United would break in and want to know what time to expect clearance and the guy would be jolted back to reality."

But money. The biggest hammer the world has to destroy companies with is money. You've got to bend, you've got to be a little bit vicious and tough if you're going to compete, a lot vicious and tough if you're going to be on top of the heap. East Island Airways did not choose to be either. That first summer the airline earned $2148 in passenger fares. It paid out $6529 in operating expenses. It lost, then, $4381.

That is a sign of disaster and despair, if and only if the first purpose of the company is to make money. But the whole outside world, all those business facts of life had to kind of gnash their teeth helplessly. Because East Island Airways is not run on the world's terms, it is run on its own terms.

"I talked to Maury about it, my lawyer," said Mrs. Friede, "and he said, 'You're not going to—this is a crazy investment and I hope you're not going into it as an investment for profit.' But he said, 'Look. You don't spend any money in the night clubs, you know, everybody needs his thing and if it's an airplane, all right. You're in a position where you can spend some money to have fun, and if this is your way, then go ahead, with my blessing. I envy you.' " She smiles a perfect, calm, world-defeating smile. "Profit was never the motive, thank heavens, but fun was, and in that way it was a big success. I really love the Bomber."

Fun. When your first motive is fun, and money comes second or third, it's pretty hard for the world to pull you down.

When destruction-through-money didn't work, the world turned to operating problems. Weather. Maintenance. Traffic delays.

"I remember a time when I was late," said Kramer. "There was a thunderstorm had closed LaGuardia and everybody else canceled air taxi flying for the night. I was at Republic Field on Long Island, and Eleanor and the passengers were at LaGuardia waiting for me. I was calling LaGuardia every hour, trying to coax the controller into saying there wouldn't be an hour delay, landing. All I had was one dry cheese cracker, waiting at Republic, and I finally got through and landed at LaGuardia and they were having a *party* in there! One guy had gone out and bought a whole delicatessen and put it in a box and dragged it over to the airport. I walk in and the guy says, 'Here, want some roast beef?' and he gave me a thing of roast beef and I had had that one cheese cracker all that time and I said, 'We're leaving, now. The airplane is leaving, now.' Back with the baggage and in they went, but their little party went right on. I said, 'Quiet, please,' I gave Eleanor a dirty look and everybody quieted down."

"He gave me a lot of dirty looks from time to time," Mrs. Friede said, "and I knew which ones were for real. He'd put up with a lot of noise and nonsense in that big back cabin, as long as it didn't interfere with his flying. But if a passenger got careless with a cigarette—well, we had a message going and we'd cool it, then."

In a way, the curious hard world finally won. When the air taxi insurance rates doubled, from fifteen hundred dollars for a summer to three thousand dollars, it was too much. But the partners don't sound beaten at all.

"I don't think we'll be running the Bomber again this summer as a commuting deal," Kramer said. "I might have to take a job someplace. But every once in a while it will come flitting into LaGuardia making that bopping and croaking noise that it makes as it taxis around, that the line boys know right away. They say to me things like, I come in at night and they say, 'My, you know, that—see them *flames* come out the exhausts!' And that noise . . . MMRRrrrowCHKkrelchkAUM . . . croaking and everything and they say, 'Boy, that's nice!' It seems to make everybody happy, wherever it goes.

"And the future? I think it would do Cessna no harm to promote one of the truly great airplanes that it built. It would be sort of a thing for them to say, 'Here is a thirty-year-old Bamboo Bomber that has just flown around the world.' So I would like to take it around the world. Because the airplane deserves to go around the world."

One has the strangest feeling that Kramer will somehow do just as he says, although the airline probably won't make a cent in profit and might even lose money, on the flight.

But that is the story of East Island Airways. You may go ahead and jump from that ledge now, if you wish. I just thought that you should know that these two people discovered that an alternate to jumping is a laugh, and a decision to live by their own values instead of the world's. They made their own reality, instead of suffering in somebody else's. According to East Island Airways, the hard earth was not made for leaping into, but for flying around.

And that bopping croaking sound you hear in the night is the Bamboo Bomber, thirty years old, taxiing for another takeoff into its adventures, blue flames from the exhaust, chuckling and chortling, and not particularly caring whether or not the world happens to approve.

N68072

A gospel according to Sam

An old guru surely must have said it to a disciple ten thousand years ago. "You know, Sam, there will never live anyone who will ever own anything more than his own thoughts. Not people, not places, not things will we ever keep for possessions through vast times. Walk a little while with them we can, but soon or late we'll each take our own true possession—what we've learned, how we think—and go separately around our lonely turnings."

"Ah, so," Sam must have said, and written it all on lotus bark.

What was it then, these thousands of years after that truth was written, that I should feel sad, signing papers to trade away a biplane that had become a part of my life? There was no question that it had to be done. My new home is edged on three sides by water, on the fourth by a high-density area. The airport, without a "control tower," thanks be, is nevertheless all hard-surface runway, is all buttered glass for the biplane to land upon, concrete strips poured into a jungle of oak trees without a single field for landing should an engine fail on takeoff. I moved nine hundred miles from the place where the biplane was at home, and the longer I left her in the hangar the worse it was; she fell to the mercy of house-hunting sparrows and cord-hungry mice. There was no choice, if I loved that airplane and wished her to live in the sky, but to trade her to someone who would fly her well and often. Why was the moment that I signed the papers such a sad moment?

Perhaps because I remembered the six years we had

flown together. I remembered the dawn in Louisiana when everything went suddenly wrong, when all of a moment she had to fly after an impossible hundred-foot ground roll, or be torn apart by a dike of earth. She flew. She had never lifted off so quickly before, she never did since, but it happened that one time—she touched the dike and flew.

I remembered the day of the handkerchief pickup, barnstorming in Wisconsin, when I had flown her hard into solid ground that I thought was only grass, slamming her propeller a hundred miles per hour into the dirt, smashing a wing, tearing a wheel loose. She didn't crush into a ball; that instant, she bounced off the ground, turned into the wind, and eased down into the shortest, softest landing we ever made together. Twenty-five times the propeller blades hit the ground, and instead of flipping on her back or cartwheeling to splinters, the biplane bounced and flew to that marshmallow featherdown landing.

I remember the hundreds of passengers we had flown from cow pastures, who had never in their lives seen a farm from the air until the biplane and I came along to give them the chance, three dollars the ride.

It was sad to part with that airplane, in spite of knowing that one never owns anything, because that flying was finished for me now, because a full good time of my life was finished and done.

The airplane I took in trade is an 85-horsepower Clip-Wing Cub. A completely different personality from the biplane, light as thirty feet of spruce-framed Dacron; that doesn't even blink at the concrete; that lifts me from take-off to a thousand feet over the trees in one length of the runway. It flies aerobatics happily that the biplane never could honestly enjoy.

Still it was all rationalization, still I felt a gray melancholy, a wistful sadness that the biplane and I had parted and that it was my fault.

It happened one day, after a practice of slow rolls out over the sea, that I realized a simple fact that most people discover who have to sell an airplane. I learned the fact that every aircraft is two separate life-forms, not just one. The objective frame, the steel and spars, is one airplane. But the subjective, the airplane with which adventures have been shared, with which we forge this intense personal bond, is another machine entirely. This machine, flying, is

our breathing past, is as truly our own as thought itself. It can't be sold. The man whose name is now on the registration papers of the biplane does not own the biplane that I do, that one hushing down through dusk to a summer hayfield in Cook, Nebraska, wind sighing in her wires, engine a soft windmill, gliding over the road at the edge of the hay. He doesn't own the sound of Iowa fog changing to raindrops on the top wings, pomming down on the drum-cloth of the lower wings to wake me, asleep by the ashes of last night's campfire. The new owner didn't buy the delight-terror cries of young lady passengers at Queen City, Missouri, at Ferris, Illinois, at Seneca, Kansas, who found that steep turns in an old biplane feel the same as stepping off the roof of the barn.

That biplane will always be mine. He will always keep his own Cub. I learned this from the sky as well as Sam from his guru, and it was no longer necessary to be sad.

A lady from Pecatonica

Remember when you were a kid, how important it was to be loved and admired? How great it was, now and then, to turn up the hero of the game, with the girls watching and the other guys glad because you scored a point or brought glory upon the team? A strange thing, flying, to come along and reverse all that.

I was barnstorming Pecatonica, Illinois, in the summer of 1966. It had been a good weekday, we had flown thirty passengers from supper to sunset, and there was time for just one more flight before it would be too dark to fly. The crowd was still there, parked in cars or standing in groups of friends, watching our planes.

I stood at the wing of my biplane and called to them in the twilight. "One more ride, folks; last ride of the day—best ride of the day, coming up right now! No extra cost, just three dollars! Room for only two passengers!"

Nobody moved.

"Look at that sunset, all red up there! Twice as pretty when you see it from the sky itself! Step into this cockpit and you can be right in the middle of it all!"

The hills and trees were already dark silhouettes on the horizon, like the cutouts along a planetarium rim before the lights go down for the star show.

But nobody wanted to fly. I was helpless—the keeper of a magnificent beautiful gift, trying to share it with a world that wasn't interested.

I tried one more time to convince them, and gave up. I started the engine and took off to see the sunset all by myself.

It was one of those startling times when I hadn't known how truly I had spoken. The ground haze topped out at fifteen hundred feet, and from the crystal air above, in the last of the sun, it was a sea of liquid deep gold, with the hilltops rising green-velvet islands out of that sea. It was a sight that I had never seen so purely, and the biplane and I climbed alone, watching, soaked in the color of that living time.

Somewhere around four thousand feet we stopped our climb, unable to take the moment all so passively. The nose came up, the right wings went down, and we fell away in a power-off wing-over that melted into a loop that eased into a barrel roll, the silver propeller just a slow fan whisking away in front of us as we came down, earth beneath us, earth over our head. It was flying for the pure joy of being in the air, and for thanks to the God-symbol sky for being so kind to us both. We thought humble and proud at the same time, all at once in love again with this painful bittersweet lovely thing called flight.

The clear wind streamed around us in that airy shriek it has at the bottom of loops and rolls and then it went soft and calm, gently flowing over us at the tops of our great lazy hammerhead turns when we almost stopped in the sky.

The biplane and I, we who had shared so many adventures—storms and sun, rough times and smooth, good flights and bad—plunged at last together into that pure golden sea. We sank far down into it, wings going level, and we glided to the bottom, to land on the dark grass.

Switch off, and the propeller clanked sadly around to a stop. I stayed for one long minute in the cockpit, not even unbuckling the parachute. It was very quiet, although the crowd was still there. The high sunlight must have been flashing from our wings, and they must have stayed to watch.

Then in the middle of that silence I heard one woman say to another, her words loud across the night air:

"He has the courage of ten men, to fly that old crate!"

It was like being slugged with an iron pipe.

Oh, yes, I was the hero. I was loved and admired. I was the center of attention. And I was disgusted, instantly, with every bit of it, and with her, and I was terribly deeply sorry. Woman. Can't you see? Can't you even begin to know?

So it was in Pecatonica, Illinois, in the summer of 1966, in the cockpit of a biplane just landed, that I found it is not being loved and admired by other people that brings joy to living. Joy comes in being able, myself, to love and admire whatever I find that is rare and good and beautiful—in my sky, in my friends, in the touch and the soul of my own living biplane.

". . . the courage of ten men," she had said, "to fly . . . that old . . . crate . . ."

There's something the matter with seagulls

I've always envied the seagull. He seems so free and uninhibited in his flying. In contrast with him I fuss and figure and clutter up the sky with noise just to stay in the air. He's the artist. I'm the tyro.

Lately, though, I've begun to wonder about the gull. Although he zooms and dives and turns with a grace that leaves me green-eyed, that's all he does—zoom and dive and turn. No aerobatics! Either he lacks initiative or he's faint-hearted. Neither of these conditions is becoming to a top hand in the air. I don't want to be hard on him—don't expect eight-point rolls and clover-leafs initially but it doesn't seem too much to ask for a simple loop or an easy slow roll.

Many times as a confirmed gull watcher, I've been sure some young ace was going to show me something. He'd come screaming down toward the water, building up speed enough to satisfy any pilot and pull up . . . up . . . up . . . till I would be sure he was going over the top. I'd stand there muttering "Pull it in!" but something always seemed to happen. You could see him slackening off the G's and the pullup arc would widen. He would roll out and lose himself in the crowd of his fellows as if thoroughly ashamed that he had dogged it.

"You look so lordly," I'd think, "but put a sparrow on your tail and I'll bet you couldn't shake him."

Other birds have developed some precision flying and a few aerobatics. Geese sometimes fly a passable formation, and that's worth mention. Some geese, though, evidently

fear the mid-air collision. Many a formation has been spoiled by number four or five taking too much spacing and straggling all over the sky. Add to this the quacking of the others telling him to close it up, and it's just plain sloppy flying. No wonder hunters shoot them down.

The unlikely pelican is almost a candidate in the aerobatic field. He can execute a neat split-S, but he doesn't meet a prime requirement of the maneuver: pulling out. He doesn't even seem to try to pull out, and ends in a geyser of white spray in the water. This isn't even playing the game.

So we come back to the seagull. We can excuse pelicans and geese, robins and wrens, but a seagull was plainly designed for aerobatics. Consider these qualifications:

1. Strong wings and spars properly proportioned.
2. Slightly unstable design.
3. High limiting Mach.
4. Low stall speed.
5. Rugged construction.
6. Extreme maneuverability.

But all these factors are useless because he isn't aggressive in his flying. He's content to fly his life away practicing fundamentals that he learned during his first five hours in the air. So, although I do admire the seagull and the free way he flies, if I had to forego an aggressive spirit to trade places with him, I'd choose my noisy cockpit any day.

Help I am a prisoner in a state of mind

Something must have gone wrong at the very first, when I was learning to fly. I remember that I had a very difficult time believing that these little machines actually lift up *off the ground;* that one minute they are all solid on the earth like a pool table or an automobile or a bright-fabric hot-dog stand and the next minute they are *in the air,* and you can stand beneath them at the airport fence and they go right over you and there is nothing at all connecting them to the ground, nothing there at all.

It was hard to grasp that, to take it in. I'd walk around an airplane, touch it, knock on it, rock it a little bit by the wingtip, and it merely stood there: See, student? Nothing up my sleeves. No gimmicks, no tricks, no hidden wires. It's real magic, student. I happen to be able to fly.

I couldn't believe it. Maybe I still don't believe it, today. But the point is that there was something spooky going on, eerie and mystical and otherworldly, and maybe that's how I got myself walled into this corner and now I'm trapped in here and can't get out.

It's all gone worse, there's nothing about flying that I can take for granted, there is nothing there that is common and everyday. I cannot just drive down to the airport and step into my airplane and start the engine and take off and navigate somewhere and land and let it go at that. I would like very much to do this. I want to do this desperately. I envy the pilots I see who casually hop into their machines and go flying out on business or charters or instructing, or the ones who fly for sport and don't have to make such a big thing about it. But I'm a prisoner in this state of mind that sees flight so all-fired awesome and cosmic that it won't let me do the simplest things at an airport without insisting that stars are changed in their course because I do them.

Like . . . look. I come out to fly, and before I'm even out of the car, before I'm even in sight of the field, I look at the sign AIRPORT and it sets me off. AIRPORT. A *port of the air,* as a seaport is a port of the sea . . . And I think about the little ships of the air sailing through the sky to this one port of all the possible ports that they could go to, choosing this one place, now, to return to earth. Touching down at this island of grass that has been waiting all special and patient for them here, and then taxiing in and being tied to their moorings, rocking gently in the wind as little ships rock in their harbors.

I'm not even *there* yet, I'm just seeing the airport sign, and maybe a Cessna 172 off in the distance, hushing down final approach, disappearing behind the trees of the roadside foreground onto what I know is a large level place for landing. Where did that Cessna come from, and where is he going? What storms and adventures has that pilot faced in his time, and his airplane? Perhaps many adventures, perhaps few, but they've been out in that vast tremendous sky somewhere, and they've been changed by it and now they've come from it to this one little harbor, the very port of the air that I'll see the moment I turn this last corner.

I can't just say "airport," like that, simply, and go on to the rest of a sentence. It is always "airport . . . airport . . ." and I get going on it that way and either make the wrong turn or run off the road or frighten some innocent person pulling out of the gas station. It's such an exciting place, an airport, that if I dare to stop and think about it or even use the word, the chances of having a routine flight are pretty well out the window before I even stop the car.

Car stopped at last, though, having avoided collisions-by-daydream with the thousands of things they put along roadsides to collide with, the first thing I see is my little airplane, waiting for me. And I can't believe it . . . That is an AIRPLANE, and it belongs to ME! Incredible. All those special-formed pieces and parts fitted so carefully together into such a beautiful clean sculpture, they can't be mine! An airplane is a thing too beautiful to own, like the moon or the sun. There's so much there! Look at the curve of that wing, the sweep of the fuselage into the vertical stabilizer, the sparkle of the glass and glint of the sun on metal and fabric . . . why, that belongs in the main gallery at the Museum of Modern Art!

So what, if I worked my heart out for the money to buy it, or if I rebuilt the whole thing up from sticks in my basement, or if I cared for it to the exclusion of every other element of a normal life. So what, if I spend nothing on liquor or cigarettes or movies or bowling or golf or boating or eating out or new cars or stocks or savings. So what, if I value this airplane when no one else in all the world has valued it, it still makes no difference, it's still unbelievable that anything so beautiful could happen in the world to make an airplane mine.

I get to thinking about this, looking at the instruments and the radio, touching the control wheel, the fuel selector, the navigation-light switches, the upholstery of the seats, the little numbers on the airspeed indicator and the way the altimeter needle moves when I turn the adjusting knob, listen to the wind sliding ever so gently over the grass and around the curves of the airplane, and half an hour's gone past, whap, like that. I sit there all alone in the airplane on the ground without moving much or saying a word, just looking at it and touching it and thinking about the thing and what it can do, that it can fly, and half an hour is half a second, it's gone before one tick of the aircraft clock.

It can fly. Anywhere. And I know just what to do with my hands and feet on all the knobs and controls and pedals, in just which order, to make the airplane come alive and actually leave the earth and point anywhere on the globe, anywhere at all, and if I really want to get there, get there. Anywhere. From right where I'm sitting this moment. In this airplane. New York. Los Angeles. Canada. Brazil. France, if I want to install an auxiliary fuel tank, and then Italy and Greece and Bahrein and Calcutta and Australia and New Zealand. Anywhere. It is so very hard to believe that, and yet it is true without the shadow of question by anyone who flies airplanes. Anybody else can take it as fact proved a thousand thousand times over; I sit there in the cockpit and another half hour goes tick on the clock and it is impossible to believe. I understand it, all right, but I cannot honestly say that I can grasp it, that I can believe it, all at once, that an airplane can fly.

That's just a start, that's not even getting off the ground. That one word "airplane" means so much! How can anyone not like an airplane, or fear it, or find it less than beautiful to behold? I'm unable to accept that there can be a person alive, any human being anywhere, who can look upon this creature of curves and wings and walk away untouched.

The time comes, eventually, when I can force myself to get the engine going, the propeller turning, but I tell you it takes superhuman concentration to do it. Because I reach down to that handle and it says on it STARTER. Starter. That which starts, that which begins the whole journey up into the sky, across any horizon in the world. Starter. Touch that and my whole life changes again, events are set in motion that otherwise would never happen. Sounds will sound on the planet when otherwise there would have been silence; winds will twist and blast when otherwise there would have been calms; motions and blurs when otherwise there would have been still sharpness. Starter. It is so momentous that I sit there, hand suspended in mid-reach toward it and I have to swallow and tremble and ask whether I am man enough, whether I have the divine Permission of God to set all these galaxy-changing events in motion. The handle waits there, and the word on it is STARTER, all

right, black letters on ivory plastic, letters faded from being touched so often over the years.

Touch that handle and a whole separate cosm lives: the engine. ENGINE. Dead cold steel for now, but a moment from now, if I will it, warm life and oiled bearings turning round and sparks flickering in darkness and pulses through eel-black wires and gages lifting awake and smokes and thunders and purrs and the standing whirlpool of sparkle and wind that is the propeller. PROPELLER. It propels. Forward. Into what? Into spaces that have never felt the touch of man, into events that test us all, against which we can measure our worth as human beings at work on destiny . . .

You can see the kind of trap I'm caught in. I can't do the simplest work at the airport (oh, port of the air, haven of the little arks that sail the skies), I can't just step into the airplane (wonder-filled machine built of magic prin . . .) and start (to set in mo . . .) the darn eng . . . (cosm . . .) without all the world roaring out in great golden glory-streaks and trumpets sounding in the heavens and angels flapping around the clouds and singing Alleluia in chorus twenty thousand strong, man-angels with low voices and woman-angels with high voices and all so grand and magnificent that there are tears in my eyes and I'm all melted in joy and praise and gratitude to the Mind of the Universe and I haven't even touched the starter yet!

It's that way with everything aeronautical, nothing's immune, nothing that has anything to do with flying. If I stop the slightest instant over takeoff, for instance, I'm lost again. TAKEOFF. The taking off of those shackles and chains that have bound our fathers' fathers' fathers to the ground for centuries compounded, that held the woolly mammoth on the ground before them and the stegosaurs before them and the trees and rocks before them. It is our power, right now, to strike those shackles, to line up there on the end of a runway and press that throttle forward and move slow first, and faster and faster and lift the nose and clankrattleclinksnap the chains are gone. We can do this. We can take off. Any time we want, we can fly.

Or airspeed. A simple basic thought like AIRSPEED and I'm out there in the wind and my arms are wings and I can feel that air, that speed, that airspeed lifting me up, way up over the clouds away from everything false and into everything true, the clean pure straight honest sky. And there's those trumpets again, and those blasted angels, singing about airspeed. A hundred miles per hour on the dial, why can't it be a simple fact, and let it go at that? But no, never, not a chance. Got to be the glory.

You see how it is, then. Hangar. Fuel. Oil Pressure. Runway. Wing. Lift. Climb. Altitude. Wind. Sky. Clouds. Airway. Turn. Stall. Glide. Even Airline, and Flight Service, on and on and on. You see how it's got me like a rat in a trap.

It would be all right, and I've been quiet about this for a long time, because if my role is to be a martyr, I'll accept it humbly and upon my back bear the burden of this rare malady for the sake of all of those who fly.

But I speak out now because I've seen other pilots land, once in a while, and stop their engines and then stay in their airplanes longer than is necessary to fill out their logbooks, almost as if they were aware of glories. And yester-

day I met a man who confessed aloud that he goes to the airport a half hour early, sometimes, and he gets into his Cherokee 180 and he just sits there in the cockpit for the fun of it for a while before he even starts the engine and taxies out to fly.

I was delighted to meet the fellow. Because I'm going to let him be the martyr now, and not me. I won't have to bear that terrible burden anymore, or listen to those angels.

I'll just go out to my airplane and I'll climb into the thing and I'll reach out for that starter and I'll just reach right . . . out . . . for that . . . starter . . . Hm. The starter is really a beautiful creation, when you take a minute to think about it. What is it really starting, you know? It kind of makes you wonder . . .

Why you need an airplane . . . and how to get it

If you fly airplanes you've probably felt it at long intervals, when a chance coincidence led you through a particularly memorable flight, or to a specially welcome haven in a storm, or to meet a friend you might otherwise not have met, who knew something about flying that you needed to know. If such things have happened as often to you as they have to some, you might be one who believes that there is a kind of principle of the sky, a spirit of flight that calls to certain among mankind as the wilderness calls to some and the sea to others.

If you do not yet fly, perhaps you've felt that spirit when you suddenly realize that you are the only one in the street who looks up to watch an airplane overhead, the only one who slows and sometimes even stops at an airport to watch the little iron birds come down to earth and to lift off again into thin air. If you act this way, it's possible that in flight you'll find much to learn of yourself and of the path of your life on this planet.

If you are indeed one of these people, it is not coincidence that has brought you to this page or to flying. Flight, to you, is a required essential tool in your mission of becoming a human being. Here's a rough sketch of most of the people who fly, and if you stop and watch airplanes, it's a rough sketch of you, too.

Flyers are distressed when they must blindly trust uncaring others to take them where they want to go. Railroads, buses, airlines can all break down, can all be delayed and strand people in unwanted places. Automobiles only travel where highways go, and highways are lined with billboards. Flyers choose to be in command of any moving machine, and to pick the course it will follow.

Flyers feel a certain kinship with the sight of the earth unencrusted by humanity, they want to see it that way in one sweeping view, in reassurance that nature still exists on her own, without a chain-link fence to hold her.

Flyers value the fact that one cannot give excuses to the sky, that in the air it is not talking that matters, but knowing and acting. There is a person within each of them who stands off and watches how they act and fly, notices when they're happy and what they do about that. The person within cannot be fooled or lied to, and the flyer is quietly glad that the inner observer most often judges him an acceptable controlled human being.

Flyers have a sense of adventures yet to come, instead of dimly recalling adventures of long ago as the only moments in which they truly lived.

Other points in common are details: flyers have weekend horizons measured not in tens of miles, but in hundreds; they sometimes use their airplanes to aid their business; they find perspective in the air after a pressured week on the ground.

The one lasting basic current among flyers is that this act of flight is the path that each has chosen, that each needs to demonstrate his control of space and time in his own life. If you share that current, your distant wish to one day own an airplane is not idle dreaming, it is a requirement of your life, which you ignore, some radical aviators say, at the cost of your humanity.

There is another being within us all, though, who is not our friend, who would gladly see us destroyed. His is the voice that says, "Step in front of the train, leap off the bridge, just for curiosity, just jump . . ." For those born to fly, the same voice speaks different words: "Forget flying. You can't possibly afford an airplane. Be practical, after all. Let's keep our feet on the ground. What do you know about airplanes, anyway?"

It is a cautious, conservative figure, and true—ninety percent of the people who own light airplanes today can't afford to own them. They need the money for home and family, for savings and investment and insurance. But each of them one day decided that owning an airplane was more important than any other money-needing cause. To fly, for them, is an important part of home and family, to fly is itself savings and investment and insurance.

The most critical time in the purchase of an airplane is the instant in which the decision is made to own one. The crucial moment is making that decision, is setting top priority on finding an airplane. Everything else is inevitable. Not time, not money, not geography can stand in the way, because buying an airplane is almost entirely a mental action, a process uncanny to live or to watch. Decision made, the more you hold the airplane in your thought, the more you see it begin to appear in your life, as well. You don't find your airplane as much as your airplane finds you.

Once you know that you need it, the process moves quickly and automatically. What kind of airplane? New or old? High-wing or low? Two-place or four? Complex or simple? Fabric or metal? Nosewheel or tailwheel? Rugged or dainty? Fast or slow? Answer the questions, and there are the first vibrations of your aircraft, surrounding you. Your airplane has changed from wish into books and magazines with articles about different kinds of airplanes, it has changed into newspaper clippings and the famous yellow *Trade-A-Plane* from Crossville, Tennessee, with its listing of thousands of aircraft for sale and trade across the country.

Choice made, whether simple eight-hundred-dollar Taylorcraft or thirty-thousand-dollar Beechcraft crowded with radios and instruments, the airplane often appears next in miniature before it springs full-size.

One flyer decided to buy an airplane at a moment when his total bank account was less than ten dollars. He decided that he would one day own a classic little 1946 Piper Cub; fabric-covered, high-wing, two-seat simple light tailwheel airplane. Cub prices ranged from eight hundred to twenty-two hundred dollars. He held the airplane in his thought, watched it there often and affectionately.

He spent ninety-eight cents for a stick-and-paper model of the airplane (it came to $1.01 with tax), which he built in two evenings and hung from a string from the ceiling. It had entered his life in miniature, where it turned this way and that in every light breeze.

He read *Trade-A-Plane,* he spent weekends at airports, he talked about Cubs with mechanics and pilots, he looked at Cubs, and he touched Cubs. The model turned in the air. Then the strangest thing happened.

A friend of his had been given five hundred dollars to rent an airplane for company business, and mentioned it to the flyer. Knowing from his weekends a thousand-dollar Cub for sale, the flyer borrowed five hundred dollars from one friend, joined it with the five hundred dollars rental of the other, bought the Cub, and loaned it till the company business was done. Business finished, debt eventually repaid, he is now the owner of a full-size, flying, 1946 Piper Cub. As well as a tiny Cub that still hangs from his ceiling.

Another man chose a Cessna 140 for his target. There was one particularly handsome 140 at an airport nearby, but he didn't have the three thousand dollars it was worth and even at that the owner did not wish to sell the airplane. But this man so wanted a 140, he so enjoyed the personality of this particular machine, that he asked the owner if he might polish the Cessna, just to be near it. The owner laughed and bought him a can of wax.

Now polishing an all-metal airplane is no simple task, but a fresh shining Cessna 140 is indeed a lovely thing. It was only right that the owner offer the polisher a ride in the plane for payment. They became acquaintances, then friends, and today are partners in that same polished Cessna.

Everyone who owns a light airplane today at one time went through the same course: Decision, Study, Search, Discovery, and eventually it happened that they came to own all or part of the airplane they now fly.

Be extremely aware, owners will advise, be on the lookout for coincidences, for what seem to be chance events falling across your path. Coincidence is the touch of that

strange invisible spirit of the sky, which perhaps has been calling you ever so gently all your life long.

A woman pilot, dismayed by the problems of scheduling rental aircraft, decided to buy her own plane. She decided that this was important enough to spend her savings on, that flying had a higher priority in her life than money waiting in a bank. She looked at scores of airplanes, on paper and in person, but never picked just the type she wanted, though she did narrow it to something that would be all-metal and two-place. But nothing was right, she wasn't drawn emotionally to any airplane she had seen through her search, not a single for-sale sign held her eye.

Then one Saturday, just as she was leaving an airport, a white Luscombe Silvaire hushed down to land and taxied to stop near the restaurant. She liked that airplane. There was something about it that felt right to her, and though there was no for-sale sign on it, she asked the owner if he would by any chance consider selling his airplane.

"As a matter of fact," he said, "I *am* kind of looking for a bigger airplane. Luscombe's a fine flying machine, but it only carries two pepole. Yes, I might consider selling it . . ."

The woman flew the white airplane, liked it all the more for the way it acted in the sky, and knew that this was the one for which she had been searching. It took some careful arranging, an agreement to let the former owner use the aircraft until he had found his four-place, but the Luscombe was hers.

Consider. Had she not come to that particular airport at that particular time on that particular day so that she would, preparing to leave, have seen the Luscombe fly down that particular final approach to land, she would have missed it. Had the wind been from the opposite direction, she would not have seen it land. Had the owner delayed two minutes longer any time that day before he flew to the airport for a cup of coffee, she would not have seen it.

But all those things happened. The chain of peculiar coincidence that is the mark of that spirit calling us and guiding us where we best may learn, happened, and today the woman flies an all-white Luscombe Silvaire which she needs and loves.

"My work takes from me all week," she says. "My airplane gives me back on weekends."

Listen, as you search for your airplane, for the following words: "Oh, no. You don't want that airplane. You won't even find that kind of airplane around here at all." The words mean that you are getting very close. I heard just those words about Fairchild 24s a week before I found my Fairchild 24. I heard them years later when I asked about trading the Fairchild for a biplane, shortly before I traded the Fairchild for a biplane. Remember, ". . . haven't got a chance" means, ". . . you're practically standing on it."

The whole trick in searching is simply to do your best, looking, and let the eerie old sky-spirit rig events so that if you aren't careful you'll walk into the wing of the airplane you are meant to own. The spirit cannot be put down. If you haven't yet learned to fly, and if you want to fly more than anything else, you will learn. No matter who you are or how old you are or where you live, if you wish it, you will fly. It sounds spooky, but it works.

It works even if it has to take the long way around. Nearly every new pilot today, for instance, learns to fly on modern nosewheel airplanes that are built for simple handling on the ground as well as in the air. As a result, the older tailwheel airplanes have become known as fierce unpredictable demons requiring superskills for takeoff and landing, which will groundloop and roll themselves up in crushed heaps if the pilot relaxes for a moment when the airplane is landing. Yet modern trained pilots often find themselves buying tailwheel airplanes simply because these cost so much less and perform so much better than do nosewheel airplanes. The guided path had led them face to face with demons.

Not very kind of the sky-spirit to set blocks in the way of its own special human beings. But the spirit mentions something about fears are built for overcoming, and the new pilot finds himself, because he needs an airplane, because he must have an aircraft to advance along his road of knowing, owner of a tailwheel machine about which he has heard terrible unforgiving stories.

He approaches his airplane with all the enthusiasm of a riding academy student approaching Old Dynamite in his stall. But as the rider, in no hurry, gets to know the thoughts and ways of Dynamite, discovers he's terrified of wind-blown papers and a pushover for carrots, that there are times to relax and times to be very careful, riding him, so does the pilot discover that a tailwheel airplane, properly flown, is more spirited and more fun to fly than any nosewheel machine. To see the delight in a student's eye when he finds that he can handle the Dreaded Taildragger is to understand something of what the spirit of flight had in mind all along.

If you hear that call toward the sky, as many thousands of people do, flying or not, answering or not, you are required to have an airplane to become more truly yourself than you have ever been. If you know this and do your best to learn to fly and to own that airplane, trusting that nutty spirit to arrange the impossible strange eerie coincidences for you as it has for everyone now flying, the life in flight that you must have, will be yours.

Aviation or flying? Take your pick

You look at aviation and you can't help wondering. There is so much going on all at once, and the whole thing is so foreign and complicated, and there are so many roaring individualists there, all railing at each other over tiny differences of opinion.

Why would anyone, you ask, deliberately dive into that maelstrom, just to become an airplane pilot?

At the question, the tumult stops instantly. In the dead silence, the pilots stare at you for not knowing the clearly obvious.

"Why, flying saves time, that's why," says the business pilot, at last.

"Because it's fun, and no other reason matters," says the sport pilot.

"Dummies!" says the professional pilot. "Everybody knows that this is the best way in the world to make a living!"

Then the others are at it again, all talking at once, and then shouting for your attention.

"Cargo to haul!"
"Crops to spray!"
"Places to go!"
"People to carry!"
"Deals to close!"
"Sights to see!"
"Appointments to keep!"
"Races to win!"
"Things to learn!"

They are at each other's throats once more, snarling over which part of the gold of flight gleams more brightly than any other. You can only shrug your shoulders, walk sadly off, and say, "What could I expect? They're all out of their minds."

You speak more truly than you might think. The government of pure reason departs when an airplane enters the scene. It is no secret knowledge, for instance, that a tremendous number of business airplanes are purchased because someone in the company likes airplanes and wants one around. Given the desire, it is a simple matter to justify the company's ownership of the airplane, because an airplane is also a very useful, time-saving, moneymaking business tool. But the desire came first, and then, later, the reasons were trotted out.

On the other hand, there are still some company executives whose fear of airplanes is as irrational as the affection of others, and despite time or money, saved or earned, have it clearly understood that their company will positively have nothing to do with any flying machines.

For a great many people around the world, an airplane has a special charm that time cannot dissolve, and a simple test illustrates the point. How many things are there on earth today, dear reader, that you truly and deeply want to own, with that same intense longing-to-possess that you had for that metallic blue Harley-Davidson when you just turned sixteen?

So often, as we grow, we lose the capacity to want things. Most pilots are absolutely uncaring about the kind of automobile they drive, the precise form of the house they live in, or the shape and color of the world about them. Whether or not they have or don't have any particular material thing is not of earth-shaking importance. Yet it is common to hear those very men openly hungering after one specific airplane, and to see them making huge sacrifices for it.

Rationally speaking, most pilots can't afford to own the airplanes that they do. They give up a second car, a new house, gold, bowling, and three years lunch just to keep that Cessna 140 or a used Piper Comanche waiting for them in the hangar. They want these airplanes, and they want them almost desperately. More than the Harley-Davidson.

The world of flight is a world in its youth, that is ruled by emotion and hard impulsive attachments to airplanes and ideas about airplanes. It is a world that has so many things to see and do that it hasn't had time for mature reflection about itself, and because of this, like any youth, it is none too sure of its own meaning or reason for its existence.

There is a tremendous difference, for instance, between "Aviation" and "Flying," a difference so vast that they are virtually two separate worlds, with precious little of anything in common.

Aviation, far and away the largest of the two, comprises the airplanes and airmen who have interests beyond themselves. Aviation's big advantage is the obvious one: airplanes can compress a very large distance into a very small one. If New York is just across the street from Miami, one might cross that street three or four times a week, just for the change of scenery and climate. The Aviation enthusiasts find that not only is New York just across the street, but so are Montreal, Phoenix, New Orleans, Fairbanks, and La Paz.

They find that after a very modest amount of training in the not-too-difficult mechanics of the airplane and the not-too-complicated element of the air, they can constantly feed their insatiable appetite for new sights, new sounds, for new things happening that have never happened before. Aviation offers Atlanta today, St. Thomas tomorrow, Sun Valley the next day, and Disneyland the next. In Aviation, an airplane is a clever swift traveling device that lets you have lunch in Des Moines and supper in Las Vegas. The whole planet is nothing but a great feast of delicious places for the Aviation enthusiast, and every day for as long as he lives he can savor another delicate new flavor of it.

To the Aviator, then, the faster and more comfortable his airplane, and the simpler it is to fly, the better suited it is to his use. The sky is the same sky everywhere, and it is simply the medium through which the Aviator moves to reach his destination. The sky is nothing more than a street, and no one pays any attention to the street, as long as it leads to far Xanadu.

The Flyer, however, is a different creature entirely from the Aviator. The man who is concerned with Flying isn't concerned with distant places off over the horizon, but with the sky itself; not with shrinking distance into an hour's air-

plane travel, but with the incredible machine that is the airplane itself. He moves not through distance, but through the ranges of satisfaction that come from hauling himself up into the air with complete and utter control; from knowing himself and knowing his airplane so well that he can come somewhere close to touching, in his own special and solitary way, that thing that is called perfection.

Aviation, with its airways and electronic navigation stations and humming autopilots, is a science. Flying, with its chugging biplanes and swift racers, with its aerobatics and its soaring, is an art. The Flyer, whose habitat is most often the cockpit of a tailwheel airplane, is concerned with slips and spins and forced landings from low altitude. He knows how to fly his airplane with the throttle and the cabin doors; he knows what happens when he stalls out of a skid. Every landing is a spot landing for him, and he growls if he does not touch down smoothly three-point, with his tailwheel puffing a little cloud of lime-dust from his target on the grass.

Flying prevails whenever a man and his airplane are put to a test of maximum performance. The sailplane on its thermal, trying to stay in the air longer than any other sailplane, using every particle of rising air to its best advantage, is Flying. The big war-surplus Mustangs and Bearcats, moaning four hundred miles per hour down their racing straightaways and brushing the checkered-canvas pylons on the turns, are Flying. That lonely little biplane way up high in a distant summer afternoon, practicing barrel rolls over and over and over again, is Flying. Flying, once again, is overcoming not the distance from here to Nantucket, but the distance from here to perfection.

Although he is in a very small minority, the Flyer is allowed to walk both his own world and the world of Aviation. Any Flyer can step into the cabin of any airplane and fly it anywhere that an Aviator can. He can overcome distance any time it strikes his fancy.

An Aviator, however, isn't capable of strapping himself into the cockpit of a sailplane or a racer or an aerobatic biplane and flying it well, or even flying it at all. The only way that he can do this is to enter the same long training that ironically transforms him into a Flyer by the time he has gained the skill to operate such airplanes.

Far from the relatively simple process of learning to aviate, Flying rears itself a gigantic towering mountain of unknowns to the fledgling, so that where Flyers are, one often hears the cry, "Good grief, I can never learn it all!" And of course it's true. The professional aerobatic pilot, or air racer or soaring pilot, practicing every day for years, is never caught saying, even to himself, "I know it all." If he stops flying for three days, he can feel the rust when he flies on the fourth. When he lands from his very best performance, he knows that he still has room to improve.

Bring these two worlds together in any but the same man, and sparks fly. To the distance-conquering Aviator, the Flyer is a symbol of irresponsibility, a grease-stained throwback to the days of flight before Aviation came to be; the very last person one would exhibit to the general public if one would wish Aviation to grow.

To the skill-seeking Flyer, the unskilled world of Aviation has already grown too much. The poor Aviators, he says, don't really know their airplanes when they are performing any maneuver but level flight, and they are the ones who, not caring to study their machines or the face of the sky, turn themselves daily into stall-spin statistics. They are the ones who press on into bad weather, not knowing that without the ability to fly on instruments, those clouds are just as deadly to them as pure methane gas.

"No one is so blind as the man who refuses to see," the Flyer quotes in ill-concealed distaste over any pilot who does not share his own zeal to know and to completely control any airplane he touches.

The Aviator believes that air safety is the result of proper legislation and strict enforcement of the rules. The Flyer believes that perfect safety in the air means the ability of a pilot to perfectly control his airplane; that any airplane, perfectly controlled, will never have any accident unless the pilot wishes to have one and controls the airplane into it.

The Aviator tries his level best to obey every regulation he knows. The Flyer is often airborne when regulations forbid it, yet just as often refuses to fly under other conditions that are quite legal.

The Aviator trusts that the modern engine is very well designed and will never stop running. The Flyer is convinced that any engine can fail, and he is always within gliding distance of some suitable place to land.

It is the same sky over both, the same principle keeps both men and both machines aloft, yet the two attitudes are so different as to be farther apart than miles can measure.

So the newcomer, from his very first hour in the air, is faced with a choice that must be made, though he may be unaware that he is making any choice at all. Each world has its own special joys and its own special dangers. And each has its own special kind of friendships formed, that are an important part of any life above the earth.

"Well, we defied gravity one more time." Reflected in that common after-flight saying is a hint of the tie that binds airmen together, each in his own world. Airborne, the airman is matching himself against whatever the sky has to offer. The sky and the airplane combine in a challenge, and the airman, Aviator or Flyer, has decided to accept that challenge. The far-traveling Aviator has friends of similar thought and decision all over the country; his circle of friends has a radius of a thousand miles. His counterpart, the Flyer, makes his own fierce friendships, bound as he is in a defensive minority which is convinced of the rightness of its principles.

Why fly? Ask the Aviator and he will tell you of faraway lands brought right to where you can see and touch and hear and smell and taste them. He will tell you of crystal blue seas waiting in Nassau, of the bright clattering casinos and the smooth quiet river at Reno, of the horizon-wide carpet of solid light that is Los Angeles after dark, of marlin leaping up from the ocean at Acapulco, of history-soaked villages in New England, of blazing desert sunsets as you fly down through Guadalupe Pass into El Paso, of Grand Canyon and Meteor Crater and Niagara and Grand Coulee from the air. He will urge you into his airplane, and in moments you'll be covering two hundred miles per hour to some favorite place with a magnificent view and where the chef is his special friend. Back at the airport after a night flight home, locking his airplane, he'll say, "Aviation is worth your while. More than worth your while. There is nothing like it."

Why fly? Ask the Flyer and he will pound on your door at six a.m. and whisk you to the airstrip and buckle you into the cockpit of his airplane. He will bury you deep in blue engine smoke or in the soft live silence of soaring flight; he will take the world in his hands and twist it all directions before your eyes. He will touch a machine of wood and fabric and bring it alive for you; instead of seeing speed from a cabin window, you will taste it in your mouth and feel it roaring by your goggles and watch it fraying your scarf in the wind. Instead of knowing height on the dial of an altimeter, you will see it as a tall, wide air-filled space that begins at the sky and drops right straight down to the grass. You will land in hidden meadows where no man or machine has ever been, and you'll soar upslope on a mountain ridge from which the snow sifts downwind in long misty veils.

You'll relax in soft armchairs after supper, in a room whose walls are covered with airplane pictures, and feel the thunder and shock of ideas and perfection surge like a hurricane sea over the faces of skill around you. The sea calms near sunup, and the Flyer drops you off at home in the

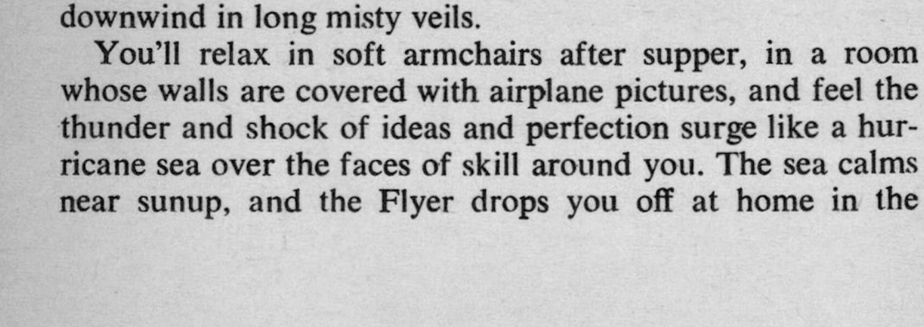

morning ready only to fall into bed and dream of airfoils and precision flying, thermal-sniffers and racing in the ground effect. Great suns roll through your sleep, and colorful checkerboard land drifts below.

When you wake you might be ready to make a decision one way or another, for Aviation or Flying.

Rare is the man who has been exposed to the intense heat of a pilot's enthusiasm, without being in some way affected by it. The only reason that this can be is the unreasonable itself, that strange distant mystique of machines that carry men through the air.

Aviation or Flying, take your choice. There is nothing in all the world quite like either one of them.

Voice in the dark

For a long time, ever since I first touched the starter switch of a flying machine, I've wanted to know what an airplane really is. A thousand hours of flying with them through good weather and not-so-good have taught me a little about airplanes, what they will do and what some of them won't do. It's taught me what goes together to make an airplane, and fairly well how it goes together. I've learned that skin is riveted to stringers which are in turn riveted to ribs and bulkheads. Mechanics have taught me that props are matched to engines and that turbine blades are fitted in balanced pairs. I've heard that some planes hold together with baling wire and others need bolts torqued to the exact inch-pound.

Through all this, though, I've never understood what an airplane really is, or why it is different from any other machine.

A few nights ago, on the anniversary of my sixth year flying airplanes, I found the answer. I walked out on the flight line of a jet fighter base and leaned against the wing of an old friend. The night was very quiet, without a moon. Dim starlight and a pair of flashing red obstacle lights outlined a black hill at the side of the runway, and I breathed JP-4 and starlight and aluminum and still night air. In the quiet I talked to my friend, who happened to be a T-33, and asked point-blank the questions I could never answer.

"What are you, airplane? What is it about you and all your wide family that has made so many men leave all they know and come to you? Why do they waste good human love and concern on you who are nothing but so many pounds of steel and aluminum and gasoline and hydraulic fluid?"

A light breeze swirled by and whistled to itself in the landing gear. As clearly as a voice in the dark the T-Bird's answer came, as if she were telling me, patiently, something she'd been telling me since we first met. "What are you," she asked, "but so many pounds of flesh and blood and air and water? Aren't you more than that?"

"Of course," I nodded in the darkness, and listened to the high lonely murmur of one of her sisters at altitude, cutting a gentle path in the quiet with her tiny airy roar.

"As you are more than your body, so I am more than my body," she said, and she was quiet again. The trim sweep of her vertical stabilizer was an intermittent silhouette against the solemn split beam of the tower beacon turning its endless path around.

She was right. As the character and life of a man is not found between the covers of an anatomy book, so the character and life of an airplane is not found in the pages of a manual on aeronautical engineering. An airplane's soul, which he can never see or touch, is something that her pilot senses: an eagerness to fly; a little bit of performance that according to the charts should not be there, but is; a spirit behind the bullet-holed mass of torn metal with three propellers feathered, touching down on an English airfield. Not the metal, but the soul of an airplane is what her pilot wants to fly, and the reason he paints the name on her cowling. And with that soul, airplanes have an immortality that you can feel when you walk onto any airport.

The air over the runways, slashed by propeller blades and burned in the Niagara-roar of a shimmering tailpipe, is part of the immortality of an airplane. The still blue lights along the taxiways at night are a part of it, and so is the anemometer at the top of the tower and the white paint that marks the runway numbers on the concrete. Even the empty sod strip at the end of a hundred miles of rolling plains lives with the calm expectancy of an engine's advancing roar and black wheels touching grass.

We can throw a DC-8 into the sky instead of a Nieuport Veestrutter, and we can throw it from a two-mile sheet of reinforced concrete instead of a muddy pasture, but the sky that the DC-8 whispers through is the same sky that held

Glenn Curtiss and Mick Mannock and Wiley Post. We can blow islands out of the sea and change pioneers' wagon trails to six-lane superhighways, but the sky is the same sky it always has been, with the same hazards and the same rewards for those who journey it.

Real flight, my friend taught me, is the spirit of an airplane lifting the spirit of its pilot into the high clean blue of the sky, where they join to share the smooth taste of joy and freedom. Like trucks and trains, airplanes have become matter-of-fact and accepted workhorses, their soul and their character aren't so easily seen as they once were. But they're still there.

Even though you cannot find an industry that doesn't benefit from flight and though there are thousands of reasons to fly airplanes, in the beginning men flew only for the sake of flying. Wilbur and Orville Wright didn't bring the powered airplane into the world to move cargo or to fight battles in the sky. They invented it for the same selfish reason that Lilienthal held tightly to his cloth-and-bamboo wings and leaped from his pyramid: they wanted to be free of the ground. That is pure flight, followed for the joy of traveling through the air as an end in itself. And every once in a while we ask, "What are you, airplane?"

Barnstorming today

When he had tightened the safety belt down over the two passengers in the front, and shut the little half-door on their leather-rimmed cockpit, Stu MacPherson held for a moment in the propeller blast by my windshield.

"You've got two first-timers, and one's a little scared."

I nodded and lowered my goggles, and pushed the throttle forward into a great roaring burst of sound and wind.

What brave people! They battle the fear of all the air-crash headlines they've read, they put their trust in an airplane nearly forty years old and a pilot they've never seen before, just so that for ten minutes they can do in fact the thing they've only done in dreams . . . fly.

The rough ground jolts hard under the wheels as we begin to roll forward . . . bit of right rudder here, and the ground is a green-felt blur beneath us . . . back on the control stick—touch it back and the thunder of the biplane moving along the ground ceases . . .

The bright sunburst biplane skims the grass tops, tearing and slicing the warm summer air with spinning propeller and cat's-cradle flying wires, and angles on up into the sky. My brave passengers look at each other in the roar of the wind and laugh.

We lift up over the grass; higher, over a field of kelly-green corn; higher still, over a wooded river lost away in mid-year Illinois. The tiny home town rests gently by the river, cooled by multiple hundreds of leafy shade trees and a faint breeze from the water. The town is an inpost of mankind. Men have been born, worked, and died here since the early nineteenth century. And there it is, nine hundred feet beneath us as we circle in the breeze, with its hotel, cafe, and gas station, its baseball game and children selling three-cent lemonade on dark-shadowed front lawns.

Worth being brave for, this view? Only the passengers can answer that. I just fly the airplane. I'm just trying to prove that a gypsy pilot, barnstorming, can exist today.

"SEE YOUR TOWN FROM THE AIR!" are our opening words to a hundred small towns. "COME UP WITH US WHERE ONLY BIRDS AND ANGELS FLY! RIDE A TESTED AND TRUE OPEN-COCKPIT BIPLANE, FEEL THAT COOL WIND THAT BLOWS WAY UP OVER TOWN! THREE DOLLARS THE FLIGHT! GUARANTEED TO BE LIKE NOTHING YOU'VE DONE BEFORE!"

From town to town we had flown, sometimes with a companion airplane, sometimes just the parachute jumper and me in our biplane. Across Wisconsin, Illinois, Iowa, Missouri, and Illinois again. County fairs, homecomings, and quiet days of quiet weeks in summer America. The cool lake towns of the north, the baked farm towns of the south; we buzzed our way, a bright dragonfly-machine carrying the promise of new views, and a chance to see away over the horizon.

More than our passengers, though, it is we who looked over the horizon, and on the other side we found that time was dead in its tracks.

Just which moment time chose to stop in the little towns of the Midwest is not easy to say. But it was clearly during a pleasant hour, in a happy time, when the minutes suddenly ceased moving one into another, when the real things ceased to change. Time stopped, I think, one day in 1929.

Those huge heavy trees of the park are there as they have always been. The bandstand, high-curbed Main Street, and the wood-frescoed, glass-front Emporium with a gold-leaf sign and a four-bladed fan stirring the air. White lap-strake churches; open-front porches in twilight; hedge clippers chopping the boundary between one house and another. The same bicycles lying right-side-down by the same gray-painted wooden doorsteps. And we found, flying, that we were part of the sameness, part of the pattern, a thread without which the fabric of town life would not have been complete. In 1929 the barnstormers chugged and clattered about the Midwest in their flake-painted, oil-throwing biplanes, landing in hayfields and on little strips of grass, di-

verting anyone ready for diversion, impressing anyone ready to be impressed.

The sound of our 1929 Wright airplane engine fitted precisely into the music of these timeless home towns. Even the same boys came out to meet us, with their same black-patch dogs running at their heels.

"Wow! That's a real airplane! Tommy, look! A real airplane!"

"What's it made out of, mister?"

"Can we sit in the pilot's seat?"

"Careful, Billy! You'll tear the canvas!"

Looks of utter awe, without a word spoken.

"Where did you come from?"

The hardest of all questions. Where did we come from? We came from where the barnstormers always come from —from somewhere out over the horizon beyond the meadow. And when we go, we'll disappear over the horizon where we always disappear.

But here we are, flying, and my two brave passengers have forgotten what a headline looks like.

Throttle back, and the engine's roar is supplanted by a polished silver fan spinning on the biplane's nose, and the whistling hush of air over wings and through streamlined wires. We circle now over the field where we will land, to view a crowd of boys, a dog, an olive-drab pile of sleeping bags and cockpit covers that is a gypsy pilot's home. Whistling, hushing, turning down across the cornfield . . . gliding quietly and bam we're down and rolling over the rough ground at fifty miles an hour, at forty, at twenty, at ten and then the black engine comes alive again to trundle us, rocking awkwardly on tall old wheels, back to where it all began. I push my goggles up on the leather helmet.

Stu is on the wing before we stop, opening the door, guiding the passengers back to solid ground. "How did you like the ride?"

A loaded question. We know how they liked the ride, how every first-timer has liked his ride since back before the clock stopped turning in the little towns of the Central States.

"Great! Nice ride, mister, thank y'." And turning away, "Lester, your house's no bigger than a corncob! Aw, it's great. Town's a lot bigger than you'd think. You can see clear on off down the road. It's really fine. Dan, you ought to give it a try."

While the engine chugs quietly and propeller blade spins easily, Stu escorts the next passengers into the front cockpit, fastens the safety belt on them, and closes the door. I put my goggles down, push the throttle forward, and a new experience begins for two more people.

Daytimes are quiet. We walk, Stu and I, through town, quiet in midday, and it's a clever electric museum. Here is Franklin's 5-10-25¢ Store, with a spring-dangled brass bell on the door and a glass front counter of rainbow candies waiting to be scooped into crackling white bags. Here are long narrow aisles floored in narrow strips of worn-down wood, and a fragrantly blended smell of cinnamon, glass, dust, and paper notebooks.

"Can I help you boys?" the proprietor says. Martin Franklin knows by name every one of the seven hundred thirty-three souls who live in this town, but it would take us twenty years to earn the same quality of his other greetings, and although our airplane from out of the past waits only a quarter mile down Maple Street, it can't make a pilot and a parachute jumper part of an Illinois town. Pilots and parachute jumpers never are, never have been, never will be part of any town.

We buy a postcard each, and a stamp, and cross the hot empty street to Al and Linda's Cafe.

We eat our hamburgers, delivered from the kitchen precisely wrapped in thin white paper, drink our milkshakes, pay our bill, and leave, not sure of the reality, but sure we had seen Al and Linda's before, maybe in some dream.

With the end of the afternoon, the worlds change. We

walk back to the dead end of Maple Street, and to our kind of reality. Here the unchanging people come to go back into time in our biplane, and from the past, look down upon the rooftops of their houses.

An unchanging summer. Clear sky in the morning, puffy clouds and distant thunderstorms in the evening. Sunsets that turn the land to misty gold, and later, fade to carbon black beneath the high glittering fireworks of the stars.

One day, we changed. We came out from the towns that never change, and tried to barnstorm and sell our rides to a city of ten thousand. The grass strip was an airport, the walls of its office covered with charts and flying rules. It wasn't the same.

It doesn't work. A biplane flying over a city is just another airplane. In a city of ten thousand, time is perking right along and we are unnoticed anachronisms. The people at the airport look at us strangely, thinking over and again that there must be something illegal about selling rides in such an old airplane.

Stu, wearing goggles and a hard hat, snaps into his parachute and lumbers into the front seat, looking as though he planned an assault on Everest instead of a quick descent to a small city in Missouri. The jump is our last hope of bringing passengers out, and our future relations with cities depend upon its success. We circle up through four thousand feet, level at forty-five hundred as the five-o'clock whistles blow around the city, signaling the end of the workday. But for us, there are no whistles. Only the constant roar of the engine and the wind while we turn onto the jump run. Stu is looking overboard absently, and I wonder what he is thinking.

He moves, and as he does an uncomfortable time begins. We usually run seventy to one hundred passenger rides between jumps, and I can't even get used to the idea of my front-seat passenger unfastening his safety belt, opening the door, and stepping out on the wing, and into the wind blast, a mile above the ground. That sort of thing just isn't done, and yet here we are with nothing but a tremendous gulf of air between wing and ground, and my friend is carefully latching the door behind him and turning to grasp a wing strut and the cockpit rim as he watches the target approach.

The biplane doesn't like these moments. It buffets and shakes heavily, its boxy streamlining broken by the awkward figure on the wing. I push hard on the right rudder pedal to hold us straight on course, and looking over my left shoulder, watch the stabilizer shudder. Mixed feelings. That's an awfully long fall, but I wish he'd hurry up and jump, to save the airplane. The airport and the city are at least beneath us. If we flew only ten percent of the people in this town, at three dollars for each person . . .

Stu jumps. The shuddering leaves the airplane. He is gone instantly, his arms spread wide in a position he calls a "cross," flicked off the wing into that big step down. As he falls, he spins, but no parachute opens.

I bank the biplane sharply and drop the nose down to follow, although he's told me he falls at one hundred twenty miles an hour and I don't have a chance of catching him now. A long time and he's still going, a black-cross silhouette roaring straight down against a background of hard green earth.

We've kidded about it before. "Stu-babe, if your chute doesn't open, I'll just keep right on flying along to the next town."

He is really smoking down. Even from directly above him, I can see that his rate of descent is fantastic. Still no chute. Something has gone wrong.

"Pull, Stu." My words are swept overboard as quickly as my friend has been. The words can do no good, will never be heard, but I can't help saying them. "Come on, kid, pull."

He's not going to do it. No main chute opens, no reserve. His body holds the same position, a little black cross spinning to the right, plummeting straight down. It is too late. I shudder with cold in the warm summer air.

In the last possible second, I see the familiar blue-and-white deployment sleeve break from the main chute pack. But too slowly, agonizingly slow. The sleeve trails out, the bright orange canopy helplessly flailing in the air, and then, quite suddenly, the chute is open and drifting serene and soft as a dandelion puff over a summer lawn.

I realize abruptly that the biplane is diving at great speed, engine roaring, wires screaming, controls stiff in the force of the wind. I ease it back into a spiral dive over the open parachute, and in half a minute have come level with it. He had room to spare . . . he's still a thousand feet above the ground!

I circle the gay canopy and the goggled jumper dangling

thirty feet beneath it. He waves, and I rock my wings in reply. Glad you made it, kid, but still, didn't you pull a little late? I should talk to him about the pull.

I hold my circle in the air while he floats on down. He flexes his knees as he always does in the final fifty feet—a last bit of calisthenics before impact. Then it seems he drops the last twenty feet very quickly, as if someone has let the air out of his canopy, and he crashes to earth, rolling as he hits. The canopy waits above him for a long moment, then settles slowly down like a huge brilliant sheet in the air.

Stu is up again at once, pulling risers and waving OK, and the jump is over.

I fly one last wing-rocking pass over him, then turn to land and fly the passengers that unfailingly flock to us after a jump.

Today there are no passengers waiting. There are a dozen automobiles lining the edge of the airport, but not one person steps forward.

Stu hastily field-packs his chute and turns to approach the cars. "Still time to fly today. Air's nice and smooth. Ready to see the town from the air?"

No.

I never fly.

Are you kidding?

We didn't bring any money.

Maybe tomorrow.

By the time he returns to the biplane I'm stretched out in the shade under the wing.

"This must not be a very air-minded town," he says.

"Win a few lose a few. You want to pull out tonight or tomorrow?"

"You're flying the airplane."

It feels funny. The city is a different place, but that's not what is strange about it. All the towns have been different.

It is a different time. Here in the city it is 1967. The year has sharp edges and angles that cut into us, that make us alien and out of our element. Traffic hums on the highway at the airport's edge. Modern airplanes come and go, all built of metal, with wide radio-filled instrument panels and powered by new and smooth-running engines.

A gypsy-pilot barnstormer can't exist in 1967, but at the same time he can. There's a big difference, some places.

"Let's get out of here."

"Where to?"

"South. Anywhere. Let's just get out of here."

Half an hour later we're up in the wind, up in the engine roar and the propeller blast. Stu is encompassed with gear, the tip of our FLY $3 FLY sign and the blue-and-white of his parachute sleeve, still field-packed, show above his cockpit rim. The sun shines in over the right side of the stabilizer, so we are heading off southeast somewhere. It doesn't make any difference where we are headed; the only thing that matters is that we're doing it now.

And all at once, there it is. Another small town of trees and church spires, a wide field to the west, a little lake. A town that we have never seen before, but one that we know in every last detail. We circle three times over the corner of Maple and Main, to see a few folks looking up and a few boys running to their bicycles. A turn west, and a moment later, propeller fanning silently around as I pull the throttle back, our old wheels whisper in the old green grass and the old ground rumbles hard beneath us.

Stu is out with the sign at once, striding to the road and the first of the curious townspeople. "SEE YOUR TOWN FROM THE AIR!"

I can hear him as I unload our sleeping bags and engine cover from the cockpits, and his voice comes clear on the clear summer air.

"COME UP WITH US WHERE ONLY BIRDS AND ANGELS FLY! GUARANTEED TO BE LIKE NOTHING YOU'VE DONE BEFORE!"

We are back where we belong. Never having been here, we are home again.

A piece of ground

There is a feeling about an airport that no other piece of ground can have. No matter what the name of the country on whose land it lies, an airport is a place you can see and touch that leads to a reality that can only be thought and felt.

Come out to the airport an hour before you fly, and just look at it, before you get involved in oil levels and elevator hinges and master switch ON. There's a row of lightplanes in their places, planes that have been there to taxi around as you roll toward the runway and take off. Look at them again. There stands a pert-nosed Cessna 140, her silver canvas windshield cover tied carefully in place. She's not just an airplane, or two thousand dollars' worth of rivets and bolts, but a man's key to relaxation and satisfaction, his way of getting up and away from the problems of people who live out their lives on the ground. Next Saturday, or perhaps every Tuesday afternoon, the windshield cover comes off and the ropes are untied. He calls "Clear!" and forgets the latest threats of nuclear war. Those, and worries about traffic tickets and W-2 forms and tie-down fees are blown back with the propwash to flatten the grass behind the tailwheel. Then he's gone and the ropes that held his airplane lie free on the ground.

Down the line near the hangar is a light twin with a company emblem on the fuselage. "You get tired of flying after the first four or five thousand hours," the gray-haired company pilot likes to say. Every once in a while, though, he'll smile a little as his bright propellers blur into life and if he hadn't denied it, you'd say he wasn't tired at all.

Look at the runway some morning when no one is flying. It lies still and quiet and is so simple: a field of asphalt. What is it then that gives a runway its mysterious, almost eerie quality of the unknown?

It is the steppingstone to flight. A runway is a constant that is found only where flight touches ground. In all the vastness of a country, in all its highways and fields and mountains and plains, flight is found only where a runway is found. The busiest city is isolated without one. The smallest farm is touched with life if there is a dirt strip out along the road. It may stand alone and deserted for weeks, but if an area on the ground can have patience, that short band of dirt has it. That time always comes when a man and his airplane seek it out, over anywhere else on earth, and come down, wheels clouding the dust, to land.

Have you ever stood in the center of a deserted runway? If you have, you know that the most striking thing about it is that it is so quiet. Airports have come to be synonyms for noise and activity, but even the runways of an international airport are frozen in silence. An engine run-up that rattles the glass in buildings on the flight line is only the bare whisper of a distant buzzing fly when it is heard from the runway. The snap of voices and radio signals exists only in airplane cabins; the runway itself takes no notice of words buried in VHF. The runway is quiet as a cathedral is quiet, and only if you listen for them can you hear sounds that come from beyond its boundary. Even the little pebbles and rocks along the edge of a runway are very special ones and—part of the world of flight—are as alien to the earth as the runway itself is alien to it.

As you stand on the broad paved field, you have at your feet the record of hundreds of landings, made in all kinds of airplanes by all kinds of pilots. The long, smooth, tapered streaks of heavy black rubber came from wheels under a man who was looking far down the runway, yet knew that beneath him the tires still had another inch and a half to descend before they touched the ground. That man has flown airplanes to ten thousand landings, and he knows many things about many places where runways lie.

Short, sudden lines of thin black abound on the asphalt surface, for on the edges of the field is a school that teaches people how to fly airplanes. These lines were made by people whose minds were crowded with the mechanics of landings, concentration on just the right amount of wing-down to offset the drift, on the stick coming back and back to hold the wheels off the ground, ready on the rudders, don't forget the carb heat on the go-around.

There is a hard set of urgent black streaks halfway down the runway, and seconds after those appeared, the air a few inches above the pavement heated in the smoke of brake discs held hard against spinning steel. There are grooves in the overrun dirt that change to hard black where they meet the runway. Just beyond the halfway mark is one curving streak that ends abruptly where the asphalt stops; the grass growing in line beyond it looks as if it has been growing there just as long as the other grass by the side of the runway, but of course it hasn't. It was once bare churned dirt beneath a cloud of grass and dust and rubber that led to the ragged tire of a war-surplus fighter.

The runway holds all this in its patient memory, along with memories of brilliant landing lights slashing the night's low clouds to throw grass shadows on the first inch of hard surface, and the sharp picture of a Waco biplane inverted at the top of a loop, its propeller an unmoving blade above the eyes of a silent watching crowd. In the memory of that runway is the cartwheeling cloud of splinters where an antique trainer landed on a broken gear.

From this place more than one boy has flown to fulfill his dream of looking down on the clouds. Beneath the dark blanket of rubber on the runway are choppy streaks from the first landing made by a gold-braided fellow who flies now as senior captain on the New York–Paris run. Out there are still the grooved stripes left long ago from the tires of a home-town boy who was last seen diving alone into combat against six enemy fighters. Whether those fighters were Spitfires or Thunderbolts or Focke-Wulf 190s makes no difference to the field of asphalt. It holds impartially the record of a brave man.

That is the runway. Without it there would be no flight school at the edge of the field, no rows of airplanes, no VHF whipping back and forth above the grass, no landing lights in the dark sky, no 140s with windshield covers neatly tied over plexiglass.

Student pilots and professionals; training planes and airliners and war planes. Men who made their mark in the air and some who made it on a hidden mountaintop. Their spirit is reflected in the beacon's majestic sweep, in the black streaks on the runway, in the roar of engines at takeoff. The spirit is held within airport boundaries from Adak to Buenos Aires and from Abbeville all the way around to Portsmouth. That spirit is the feeling about an airport that no other piece of ground can have.

Let's not practice

Practice, for her, was boring. Why, it is such grand fun just to be in the air! Look at this sky! This day! The fields all warm velvet, and the ocean . . . that's my ocean! Let's just fly for a while, and not practice slow-flying, and . . . look at that ocean!

What can you say to a student like that? It was her own airplane, her own new Aircoupe and the sky was as clear as air washed all night in rain. What can you say? I wanted to tell her, Look, you'll like a day's flying so much better, as soon as you can control your airplane with skill. Study the aircraft now, learn it well, and you won't have to think about it; later . . . it will feel like you're a pure puff of cloud, relaxed and at home in the sky.

But I could no more convince her over the sound of the engine than I could the many times I'd tried in the quiet of the ground. She was so eager to jump ahead, to plunge into the giant majesty of flight, that it was chains and hobbles for her to take one step at a time, to think about stalls and steep turns and forced-landing practice. So we flew around for a while, and I looked out at the fields and the sea and that dream-clear sky and worried about what would happen to her in this pretty day if her engine stopped.

"OK," I said at last. "Before we land, let's practice one thing. Let's pretend we're climbing out after takeoff and the engine quits, right there. Let's find out how much altitude it takes for you to turn the airplane around, correct back to the runway, and flare out for a downwind landing. OK?"

"OK," she said, but she wasn't really interested.

I demonstrated one engine-failed turn, and took one hundred fifty feet to do it, from quit to flare.

"Your turn." She botched the first one, lost four hundred feet. The second one took three hundred. The third went well, matching my one hundred fifty. But her heart wasn't in practicing engine failures, and a few minutes later we were landed and she was still talking about the lovely beautiful day.

"If you want to enjoy flying," I said, "you've got to be good at it."

"I'll be good. You know how careful I am about preflight checks. I drain all the water from all the tanks—my motor won't quit me on takeoff."

"But it can! It's happened! To me!"

"You fly those old airplanes and their motors are always stopping anyway. I have a new motor . . ." She looked at my face. "Oh, all right. Next time we'll practice some more. But wasn't this the most beautiful day of the *year?*"

Three weeks later she was solo in the Aircoupe and I was in the Swift, camera on the seat beside me as our two airplanes rolled into position for takeoff, toward the trees. It was another Tiffany day, and I had promised I'd take pictures of her airplane cruising along over the fields.

She took off first, and by the time her Aircoupe was lifting into the air, the Swift and I were rolling to follow, full throttle.

I was just breaking ground, retracting the landing gear, when I noticed that the Aircoupe was turning right, instead of left, two hundred feet in the air.

What's she doing? I thought.

The Aircoupe was no longer climbing. It was coming down, banking over the trees, the propeller a slow windmill. Without any warning, after a perfect run-up, her engine had failed on takeoff.

I was stunned, watching, helpless. She's a student pilot! It's not fair! It should have happened to *me!*

There had been no place ahead or to the side for her to land; it was all a forest of oak. Lower, she would have had no choice but the trees, but now she was turning back, trying for the airport.

She didn't have a prayer of making it all the way back to the main runway, but the cross runway might be wide enough . . .

I was a hundred feet in the air when the Aircoupe glid-

past in the opposite direction, wings gently banked, wheels clearing the last of the trees by a man's height. She looked straight ahead, concentrating on her landing.

The Swift pivoted hard around beneath me as I swung to land at once on the cross runway. I saw the Aircoupe touch the dirt at the side of the pavement, roll across the hundred feet of it and on into the cleared dirt on the other side. It took three seconds for the weak little nosegear to collapse, pitching the airplane into a spray of yellow dust, throwing the tail steep into the air, shuddering. Why couldn't it have been me?

As I rolled up, brakes smoking, the canopy of the Aircoupe slid back and she stood up in the cockpit, frowning.

I forgot to think of a fitting understatement. "Are you OK?"

"I'm all right." Her voice was calm. "But look at my poor airplane. The rpms went down and then there was just nothing. Do you think it's hurt too much?"

The propeller, cowl, and firewall were bent. "We can rebuild it." I helped her down the sloping wing from the cockpit. "That was not a bad piece of flying, by the way. You were good and slow over the last of the trees, you used every inch you had. If it wasn't for that toothpick nosegear . . ."

"Was it really all right?" The only effect of the crash was that she wanted to explain. Usually she didn't care what I knew or thought. "I wanted to turn around and land down the length of the runway, but I just wasn't high enough. When I got low, I thought I'd better level the wings and land it."

The more I stood there and looked at the space in which she had landed that airplane, the more I felt uneasy. After a minute or two looking, I began to wonder if I could have done as well, and the more I wondered, the more I doubted that I could; with all my old engines failing and off-airport landings and short-field tricks, I doubted that I could have brought that Aircoupe down any better than this student who wasted her practice time flying straight and level, looking down through the air at the fields and the sea.

"You know," I told her later, with a little more respect in my tone than I wanted to show, "that landing . . . it wasn't a bad piece of flying, at all."

"Thank you," she said.

The engine had quit from a vapor lock in the fuel line, and when we rebuilt the airplane we changed the line so

that it couldn't happen again. But I kept thinking about the way she had flown that landing. Did the practice help her, the day that we flew the three simulated engine-failures-on-takeoff? It was hard to believe that they had—she had done them only as a favor to me. I began to believe that she had the skill she needed within her all along, and the cool thinking, waiting the moment she'd want it. I thought about that, about how I had nothing at all to do with her ability to fly. Finally I came to think that maybe everything we need to know, ever, about anything, is already within us, waiting till we call for it.

I had told her so, and now she believed it: even new engines can fail on takeoff.

But I still think, now, that there are times when a flight instructor is nothing more than pleasant company when a girl wants to go flying on a pretty day.

Journey to a perfect place

The field was grass and square, a half mile through, set out in the middle of Missouri, and that's about all it was. Some hills poufed up in trees, and a pond for swimming in; way off in the distance, a dirt road and a farm, but most of all it was a soft square of green, and the color came from the dye in the cool, deep grass.

We had landed there in two airplanes, small ones, to build a fire by the pond, untie bedrolls, and cook a supper over the fire while the sun ran out.

"Hey, John," I said, "this is not too bad a place, is it?"

He was watching the final shreds of sunset, and the way that the light moved in the water.

"This is a good place," he said at last.

But strange: though this was indeed a good place to fly, we had no wish to camp more than overnight. In that short time, the field went familiar and vaguely boring. By morning, we were quite ready to take off and leave the pond and the grass and the hills to the horses.

An hour after sunrise, we were two hundred feet in the air, droning along together in loose formation over fields the color of young cornstalk and old forest and deep-plowed earth.

Bette flew the airplane for a while, concentrating on the demands of formation, and I looked down over the side and wondered if there was such a place in all the world as perfect. Maybe that's what we're really looking for, I thought, with all this flying around and gazing down from our moving mountaintops of steel and wood and cloth—

maybe we are all looking for one, single, perfect place down there on the ground, and when we find it, we will glide down to land and we will never need to fly anywhere again. Maybe pilots are just people who aren't quite happy with the places that they've found so far, and as soon as they can locate that one spot where they can be as happy on the ground as other people are, they will sell their airplanes and not go seeking any more across the sky.

Our talk about the fun of flying must be talk about the fun of escaping. Even the word "flight," after all, is a synonym for escape. Why, if I were to see, over this next little line of trees, my own perfect place, I would have no more wish to fly.

It was an uncomfortable thought, and I looked at Bette, who paid me no attention other than to smile without looking at me because she was still flying her formation.

I looked out again, and the land below changed for a moment to all the most perfect places I had seen. Instead of farmland beneath us, suddenly there was the sea, and we were turning to land on a strip cut on the edge of an ocean cliff, all lonely lost and still. Instead of farmland there was Meigs Field, ten minutes walking from the unexplored jungles of Chicago, Illinois. Instead of farmland there was Truckee-Tahoe, surrounded in razor-peak Sierra. Instead of farmland there was Canada and the Bahamas and Connecticut and Baja California, day and night, dusk and dawn, storm and calm. All of them interesting, most of them pretty, some of them beautiful. But not one perfect.

Then the farmlands were back below us and the engine power was coming in and Bette was pushing throttle to follow John and Joan Edgren's Aeronca up above the level of the first summer clouds. She turned the plane back over to me, and for a while, I nearly forgot about escape and flight and perfect places.

But not quite. Is there such a place that, found, will bring an end to a pilot's need to fly?

"Pretty clouds," Bette said, over the sound of the engine.

"Yeah."

By now, the clouds were all over the sky, puffing tall and pure up toward the sun. They had hard, clean edges, the kind you can drag a wingtip through without getting mist on your windscreen, and there were shifting, flowing snow shelves and giant cliffs and chasms all around us.

It was about that time that the answer reached out and grabbed me by the neck. Why, the sky itself is the land to which we are escaping, to which we fly!

No beer cans and empty cigarette packs strewn around a cloud, no street signs or stoplights, no bulldozers changing air to concrete. No room for anxiety, because it is always the same. No room for boredom because it is always different.

What do you know about that! I thought. Our one perfect place is the sky itself! And I looked across at the Aeronca and I laughed.

Loops, voices, and the fear of death

It was supposed to have been a simple inside loop, out off the airways, way up high, just for fun. With the wind shredding itself in a great thundering hundred-mile cry through the flying wires, I lifted the biplane's nose through a steep climb, through straight up, through an inverted climb . . . then stalled there, hanging from the seat belt upside down over thirty-two hundred feet of clear and empty air. The control stick went dead in my glove, the airplane wallowed lazily this way and that, and fell flat, like a giant slow-motion pancake, out of the sky. Dust and hay from the cockpit floor poured up past my goggles and the wind changed from clean thunder to a strange loud buffeting hum, a thirty-foot bumblebee in agony.

The nose made no particular effort to point down, the engine stopped in zero G, and for the first time in my life I was pilot of an airplane that was falling . . . just as if it had been derricked off the ground and cut loose.

I was annoyed at first, then apprehensive at the way the controls didn't respond, then I was quite suddenly afraid. Thoughts flicked through me like tracers: this thing is out of control there's altitude to bail out but my airplane will be killed this is the lousiest loop I am the worst pilot what's this *falling*, airplanes don't fall like this c'mon get that nose down . . .

Through it all, the observer behind my eyes watched with interest, not caring whether I lived or died. Another part of me was scared to panic, and cried this is not fun I don't like this at all WHAT AM I DOING HERE?

What Am I Doing Here? The question has fired itself, I'll bet, at every pilot who ever lived. When John Montgomery set himself to cut his glider loose from the balloon that carried it aloft, he must have thought, What am I doing here? When Wilbur Wright knew that he couldn't get the wings level before the Flyer hit the ground; when the test pilots discovered that the Eaglerock Bullet or the Salmson Sky-Car, after fifteen turns of a spin, would not recover; when the mail pilots, lost above a sea of fog, heard the engine die on the last of its fuel—they all heard that question from the terrified voice within them, though they may not have taken time to answer.

"Any pilot who says he's never been scared," goes the saying, "is either stupid or a liar." There are exceptions, perhaps, but not many.

For me it was spins, as I learned to fly. Bob Keech would sit calmly over there in the right seat of the Luscombe and say, "Give me a three-turn spin to the right." I'd hate him for it and go tense as steel and dread the moments ahead and bring the stick full back and force right rudder, my face dead as old soap. I'd hang on, eyes squinted to count the turns, recover at last. I'd think in pain as I leveled, I know what he's going to say. He's going to say, Now give me three turns left. And Keech would sit over there, arms folded, and say, "Now give me three turns left."

Yet that hour would fly past and we'd come skimming into the pattern and land and I'd barely set foot on the ground when my fear was forgotten and I was desperate to fly again.

What Am I Doing Here? The student on cross-country hears the question while he searches the checkpoint thirty seconds overdue. Many other pilots hear it when the weather around them turns from good to not-so-good, or when the engine misses a beat or the oil temperature turns a shade too high and the oil pressure a shade too low.

It is one thing to lean back in flight-office chairs and talk about how great it is to fly, it is another thing entirely when you are up in the air and the engine blows up and the windshield goes liquid gold in oil and the only place to land is that *little tiny* oat field down there, along the crest of the hill, with the fence at the end.

When it happened to me, there was a continuous dialogue all the way down to the ground, or, more precisely, there were two monologues. One part of me is intent on turning to final approach, holding airspeed just so, shutting off magnetos and fuel, judging the glide, steepening the bank because we are too high . . . The other part is gabbling in fright. "See? You're scared, aren't you? Big deal, you've flown all these airplanes and you think you like to fly but now you're *afraid!* First you were scared the engine was on fire, now you're scared you'll miss the field, aren't you? YOU'RE A COWARD, YOU'RE ALL BLUFF AND TALK AND YOU'RE NOT HAPPY NOW AND YOU WISH YOU WERE ON THE GROUND AND YOU ARE AFRAID!"

That day we made the landing in fairly fine style, propeller stopped, oil streaking the airplane in the strange beauty of liquids blown by the wind, and I was a proud peacock to set it down without a scratch. But even as I congratulated myself on the landing, I remembered that accusing voice telling me how scared I had been, and was distressed to admit that it had been right. Afraid or not, though, here was the machine safely landed in the oat field.

What Am I Doing Here is not supposed to have an answer. The voice that asks is hoping we'll reply without thinking, "I shouldn't be here at all. It is a mistake for man to fly and if I get out of this alive I will never be so foolish as to fly again." The voice is content only when we do nothing at all, when we are completely idle. It is the voice of paradox, of self-preservation carried to the point of death.

The way to make time pass slowly is to stay absolutely bored. Bored, minutes are months, days take years to pass. The way for us to live the longest possible life is to sit ourselves in a blank gray room, waiting for nothing, through the years. Yet there's the ideal that the voice asks us to choose—to stay in this body, this room, for as long as we can.

What Am I Doing Here has another answer, however, one we aren't supposed to find . . . I Am Living.

Remember, as a child, the challenge of the high board at the swimming pool? There came the time, after days of looking up at that board, when you finally climbed the cold wet steps to the high platform. From there it looked higher than ever, the water looked a thousand feet down. Perhaps you heard it then, What am I doing here? Why did I ever climb to this place? I want to go back where it's safe. But there were only two ways down: the steps to defeat or a dive to victory. No other choice. Stay on the board as long as you wish, but soon or late you must choose.

You stood on the edge, shivering in the hot sun, deathly afraid. At last you leaned too far forward, it was too late to retreat, and you dived off the edge. Remember that? Remember the joy that fired you back to surface so that you broke clear like a porpoise, streaming water, shouting YEE-HOO! The high board was conquered in that instant, and you spent the rest of the day climbing steps and diving down, for fun.

Climbing a thousand high boards, we live. In a thousand dives, demolishing fear, we turn into human beings.

That's the charm, that's the siren song of flight: flight is your chance, pilot, to destroy fears on a grand scale, in a high and beautiful country. The answer to every fear, be it of high board or of three-turn spin, is knowing. I know how to hold my body as I leave the board, so the water will not hurt me. I know how the wing stalls and the rudder forces it to spin. I know that the world is going to blur like a runaway green propeller and the controls will fight against my hand. I know the opposite rudder pedal will be hard to push for the recovery, but I know I can push it, and the spin will stop at once. Before too long, knowing, I climb high and do spins for fun.

It is only the unknown that is fearful. As clouds lower about us, for instance, we are unafraid if there's a runway in sight to land upon. We fear low ceilings only when the unknown lies below . . . fields or hills or treetops to come down in, when we have never once landed on field or hill or tree. But if we have landed in fields for years, if we know what to look for and how to control our airplane throughout, then landing in grass is no more frightening than landing on a mile of concrete.

All of life, some say, is a chance to conquer fear, and every fear is part of the fear of death. The student who grips the controls in apprehension is apprehensive of dying. His instructor alongside, saying, "Don't worry. Relax. See? You can take your hands off the wheel and the airplane flies smooth as a feather . . ." is proving that there is no death nearby.

Every pilot first conquered the fears of a narrow envelope of flight. We first knew our airplane and ourselves well enough only to fly around the pattern on sunny days. Then we knew more and flew into the practice area; then out into the world, then into cloud and rain, over seas and deserts—all without fear, all because we know and control the airplane and ourselves. We grew toward becoming human, and we fear only when we lose control.

We learned to avoid when we could not control, which is to say that we began to overcome stupidity. Don't Fly Through Thunderstorms is an axiom most pilots accept without testing. Never Trust Your Life To An Engine is a less heeded one, most often ignored by those who have never heard an engine stop in flight. Those pilots who fly without parachutes on black-night cross-countries and over seas of fog have no idea of where they might land if the engine quits, and without knowing haven't a prayer of controlling the crash.

It is a terrible empty feeling to have a guaranteed certified approved modern engine snap its crankshaft or swallow a valve or run out of gas when the tank reads FULL. The feeling is all the worse when one can't see to land, worse yet when one can't bail out, and reaches ultimate despair when one finds he is a trapped and helpless passenger in his own airplane.

Certainly there are hundreds of pilots who fly without fear through black nights and over miles of fog, but their peace comes not from knowing and control, it comes from blind faith in the crate of metal parts that is an engine. Their fear is not overcome, it has simply been masked by the sound of that power plant. When that sound fails in flight, I give you fear, stronger than ever. It is not legality or guarantee that determines our safety, but how well we can fly.

I've been called Daredevil for flying passengers from wide clear hayfields, Chicken for refusing to fly them from a narrow runway facing hills and trees. Wild Crazy Irresponsible for picking up handkerchiefs with a wingtip, Overcautious for deciding not to fly at night without a parachute. But still I think that fear is something to be conquered in a fair fight, not ignored or swept under a rug of illusions that engines never fail. Fear, fear—you are a demanding enemy.

The biplane fell down from the sky, wallowing, buffeting. *What am I doing here,* the voice screamed. It took a second to answer. I'm living. And I bail out if we're not flying by the time we reach two thousand feet. At two thousand feet I'll pull the seat belt release and fall free, clear the airplane, and pull the ripcord. A shame to lose it because I can't fly a simple loop. I'll never live it down.

Slowly, like a big floating safe, the nose of the biplane eased downward. The terrible throbbing buffet began barely to fade, and the airstream to smooth. Maybe . . .

We roared through two thousand feet pointing straight down, under control again, and the engine blazed once, coughed, and burst back into action. Oh boy, the voice said. You nearly had it that time and you were scared as a rat. Scared to death. This flying business is not for you, is it?

We climbed back to three thousand feet, put the nose down till the wind shredded itself in a great thundering hundred-mile cry through the flying wires, and this time with a good positive pullup we flew a fine loop, the biplane and I. Then another, and another.

What we are doing here? Overcoming the fear of death, of course. Why are we in the air? We're practicing, you might say, what it is to be alive.

The thing under the couch

The seat belts in the airplanes are different, for one thing. Instead of an American strap and buckle, there's a four-way affair that traps you in the cockpit like a fly in a spider web. The parachutes here are different, too. All the harness webs snap into a single steel block which, turned and hit hard, releases everything at once. Everyone drives about on the wrong side of the aerodrome roads, speaking in Irish brogues of petrol and carburetors, of stall turns and flick rolls instead of hammerheads and snaps, circuits instead of patterns, undercarriage instead of landing gear. It's not hard, in Ireland, to feel like a lonely foreigner.

The aerodrome is a great green square, three thousand feet on a side, grazing a block of puff-ball sheep that frighten easily but still need to be driven out of the way by a low pass before landing.

On this field one Sunday afternoon appeared a sort of Taylorcraft fitted out with an all-glass cockpit, and a little in-line engine, which I found out was an Auster. The pilot was one Billy Reardon, and the first thing he did after we met was to offer the lonely foreigner a ride in his airplane.

It was one of those parallel-world stories out of science fiction, when life feels the same as normal, but isn't, quite. The propeller turned clockwise instead of American counterclockwise; the control stick linked not to wires under the cabin floor, but to a strange yoke assembly beneath the instrument panel; the tachometer needle swung not smoothly from low rpm to high, but shuddered in quick whiplash leaps as though caught on stop-motion film.

Still, the Auster lifted up from the ground and sailed over rock walls and emerald hedges into a sky remarkably like the sky of home. We flew for twenty minutes, Billy Reardon showing the character of his airplane as would a pilot, I think, of any country. My two landings were among the worst I've ever made, but Billy stood tactfully by with an excuse that he hoped I'd believe. "She takes an hour's flying to get used to, really. She stalls at only twenty-eight miles per hour—you'll have her on the ground and along comes a little gust and you're flying again!" I liked Billy Reardon, for saying that.

Then, days later, I went to dinner at the house of Jon Hutchinson, an Englishman flying BAC-111s for Aer Lingus out of Dublin, owner of a 1930 Morane parasol, just flown after a year's rebuilding. There were photographs of airplanes on his walls, as there are on mine at home; he had shelves of aviation books, as I have.

We talked, after dinner, and all at once he said, "Let me show you . . . the most beautiful . . ." and he was on his hands and knees reaching under the living-room couch, sliding something heavy. What it was, under there, was a black steel cylinder for the two hundred thirty horsepower Salmson engine of the Morane.

"Isn't that a pretty thing?"

It shone like printer's ink, the machined cooling-fins catching light, in the room.

To whom, I thought, to how many people could he have said that, would he have admitted that there was indeed a big old engine part under the couch? Perhaps only to another citizen of his own country, of the sky. I was honored.

"Now there's a beautiful cylinder, Jon. Beautiful. What's this, here? Three spark-plug holes?"

"No, this one's for the primer . . ."

A week later I came to know another Aer Lingus pilot, who kept his Tiger Moth at the same green sheep-meadow field from which I flew. Roger Kelly's voice, except for the Dublin accent, sounded like voices I've heard more and more often in the last few years.

"The fact that you've got Air Line Transport Pilot written on your license doesn't mean you fly any better," he said. "One day these pilots who fly for the money of the job, they're going to lose everything, the cockpit's going to burst or some such thing and they'll be left with a stick and rudder and they won't know how to fly."

He didn't mean it literally, perhaps, but he did mean that distressing feeling most sport pilots have felt, a moment later: "The day they make me put a radio in the Moth is the day I give up flying."

It was about that time, I think, that I finally learned that an airman leaving the boundaries of his nation isn't a foreigner at all. In whatever part of the world he travels, chances are there's a couch under which lies a cylinder for an airplane engine; chances are there's another airman, who put it there, and finds it beautiful.

The $71,000 sleeping bag

A ferry flight, that's all, to take a Cessna Super Skymaster from the factory at Wichita to the distributor at San Francisco. Not much worth note could happen on such a routine flight, and nothing did. It happened on the ground.

The Skymaster and I had landed at Albuquerque late in the evening, taxied to the far end of the field, the west end there, the Cessna dealer. I walked to the new terminal for a bowl of soup and a bunch of crackers, and about midnight walked back to the plane.

I play-act, sometimes, when I fly a kind of airplane that I don't often fly, and pretend I'm the person I'd expect to see flying that plane. In the Skymaster, I was an executive pilot, walking back across the line to my company machine. Stereotype business pilot, all solemn: facts and figures, little briefcase, black bag full of Jepp charts—you know the kind. This was me, moving through midnight, making a note to check the weather now although I wouldn't be flying till morning. Cool. Level head. No nonsense.

But as I was going my businesslike way, just stepping up from the street onto the parking ramp, by the low cyclone fence, I happened to notice the silhouette of the Skymaster against a big floodlight . . . the twin shark tails all still and powder-black against the light. I felt a great surge of affection for the airplane, for that thing.

Just because we had come far together in one afternoon, I guess, and against headwinds.

Affection for an airplane. Somehow, I had never thought that company pilots felt that way. But they do.

That was the first thing.

Mounted on the Cessna hangar is a loudspeaker that is set to the tower frequency, and turned up way loud so the lineman can hear it and be ready to flag the in-bounds in for gas. Nothing but static, at that hour, static very highly amplified in the speaker. But then there was a burst of words, the voice of some guy flying unseen in the night. "Hello Kirtland Tower. Twin Beech niner six Baker Kilo is at the Pass, inbound to land."

No sound in the sky, just that voice in the speaker, echoing, with the throb of engines in the background.

Then a few minutes later, I heard the faintest little muffled rasp of propellers humming around, and saw the slow streak of navigation lights. The guy had taken a step into reality; he was slowly changing from one dimension into life. "Six Baker Kilo is five out on a straight-in."

"Baker Kilo cleared to land." It was a gentle drama, a play on a ten-mile stage, and I was the whole audience.

A few minutes later came the chirk-chirk of wheels touching concrete, the sigh of engines fading back from approach power. Then silence. Then the sound of engines again rumbling around at idle, louder and louder till they gasped suddenly and coasted their propellers around to a stop just fifty feet from where I stood by the Skymaster. The quiet little noises, then, of flight's end: chock scrapes, door sounds, and the talk of pilot to copilot.

That was the second thing.

When the Beech pilots had left, I put the right seat of the Skymaster to full recline, stretching out on it as best I could. Suit coat for a blanket, padded headrest for a pillow. It wasn't comfortable at all . . . not a tenth as pleasant as unrolling one's sleeping bag under the wing of a Champ and looking up at the stars.

This airplane was different. It was sheet metal instead of cloth and dope, radios and omni and ADF and DME and marker beacon and EGT and autopilot and trim and flaps and prop control and mixture instead of nothing. But the stars were the same stars.

By sunup, I was convinced that the Cessna Super Skymaster, although it is a great twin-engine plane that can never kill a pilot with the terrible yaw of engine loss in the weather at full gross, is a lousy sleeping bag. For $71,000, I thought thay should at least make the airplane a little easier to spend the night in. Then, too, I found that you don't want to hang your good shirts on the aft propeller, because you'll get exhaust powder all over them. The front prop is okay, but the man with a $71,000 airplane will certainly have a larger wardrobe than can be hung on one propeller.

That was the third thing.

At dawn we were airborne, the Cessna and I, and before noon we were landing in California. An abysmally poor sleeping bag, but not a bad machine for going places.

Machine? I thought, and saw again the shark-fin silhouette at Albuquerque, the Beech pilots brought alive, the $71,000 sleeping bag. They are all alike, if you look at them at just the right times. Old or new, rag or tin, no airplane's a machine. And what they are instead is a lot of what makes flying kind of fun.

Death in the afternoon— a story of soaring

He didn't say anything till the afternoon of that first day. Then, as we lowered ourselves into the team sailplane, strapped ourselves tightly about with webs of parachute and shoulder harness and seat belt, tested the flight controls and spoilers and towline release, he said, "It's like getting ready to be born. A baby feels this way, strapping into its new body."

I warn you. He's wont to say things like that.

"This is no body," I said, firm but not harsh. "See? Manufacturer's data plate, right here. Schweizer 1-26, single-seat sailplane. And all these others on the runway are 1-26 sailplanes, too, and this is Harris Hill and this is competition and we're out to win and don't you forget that, OK? Let's stick with the business at hand, if you don't mind."

He didn't answer. Just tugged against the straps, tightened them down, moved the flight controls light and quick, the way a pianist moved his fingers, quickly, in the last moment before the concert begins.

A Super Cub towplane taxied out in front of us, and a couple hundred feet of nylon stretched to join us for the launch. We were ready for takeoff.

"Helpless. Nothing as helpless as a sailplane on the ground."

"Yeah," I said. "You ready?"

"Let's go."

I fanned the rudder to signal the tow pilot. The Cub crept ahead, the line snaked out, tightened, our awkward beautiful Schweizer eased forward. The towplane pressed full throttle and we were on our way . . . in seconds we had aileron control, then rudder, and at last, elevators. I touched the stick back and the glider lifted clear of the runway, just a few feet clear, to make the takeoff easier for the Cub. We were flying, with the hard rush of wind about us, with the controls alive in our hand.

"We're born," he said calmly. "This is what it means to be born."

He took the controls without asking, flying clumsy at first with those great long wings, porpoising a bit behind the towplane till he got used to flying high-tow formation again. He did a fair job of it—not excellent, but not too bad. He was an average pilot, I'd say. Average low-time pilot.

Harris Hill fell away behind us. The Cub turned to follow the ridge, climbing, and though we felt some lift and perhaps could have released a minute after takeoff, we stayed meekly on tow, considering it wise to use the extra help while we had it.

"You ever noticed," he said, "how much being on tow is like growing up, like a kid growing up? While you get used to the feel of living, the towplane-mother is out ahead of you, protecting you from sinking air, getting you up to altitude. Soaring's a lot like living, don't you think?"

I sighed. He talked this way, and he ignored the fine little tricks of competition. We could turn the towplane toward our course by tugging left on the tail of the Cub, with the towrope. We could keep him from climbing too fast by tugging up on it. Tricks like that can give you a free ride another few hundred yards on course, and that can make a difference, in a meet. But he ignored everything I knew, he went on about what he knew.

"A kid can take it easy, not much pressure, not many decisions, taking his tow into the air of life. Doesn't have to worry about sink, or about finding lift for himself. Being on tow is what you call *security*."

"If you'd turn a little left . . ." I said.

"But as long as he's on tow, he's not free, there's that to consider."

I was restless to get a word in. I wanted to urge him to nudge the towplane to give us that extra push in the right direction. It's not cheating. Any pilot can do it.

"I'd just as soon be free," he said.

Before I could stop him, he pulled the towline release—

VAM!—and we were loose in the sky. The high-speed noises of tow fell off to the gentle hush of a glider at cruise.

"That wasn't too smart," I said. "You could have worked another two hundred feet out of that tow, and herded him around . . ."

"I wanted to be free," he said, as if that was some kind of an answer.

To his credit, though, he did turn directly on course, pointing the nose into the wind toward the goal, forty miles distant. It was not an easy task, an upwind goal in a 1-26. To make it worse, there was a great blue hole of dead air between us and the first cumulus across the valley.

It was going to be a long tough glide to reach them, and by then we might well be too low, we might glide in beneath their rising air. He kept the nose on course, and increased airspeed to best penetration through the sinking air. Most of the other sailplanes, I noticed, stayed around the hill after release, working the ridge lift; and waiting for a thermal to give them safe altitude for a leap across the valley. A lovely sight, they were, wheeling and soaring in the quiet of the sun. Yet all the while they circled, I knew, they were watching us, to see if our try to penetrate at once would pay off. If it did, they'd follow.

I wasn't sure what I would have done, if I were flying. It is all very romantic and daring to go blasting off on course directly after release, but if you don't make it, if the sink presses you right down to the ground, you're dead, you're out of the game. Of course, you're dead if you spend all day in the ridge lift on Harris Hill, too. The game is to reach the goal, and that takes just the right blend of bravery and caution. The others had opened with caution; my friend had chosen bravery. We flew directly away from the hill, sinking three hundred feet per minute.

"You're right," he said, reading my doubt. "Another minute in this sink and we won't be able to glide back to the hill at all. But don't you agree? Sooner or later, doesn't a man have to turn his back on the safety of towplanes and ridge lift and set off on his own, no matter what?"

"I guess."

But maybe if we had waited, some thermals would have cooked off in the valley. As it was, we could stay in the air for another five minutes, and then we'd be forced to pick a field and land. I started looking for fields, a little sullen, maybe, thinking more that we should have waited, like the others. I like soaring. I don't like to throw away what might have been a two- or three-hour flight in some seven-minute speed dash for the ground, just because this guy feels daring. Four hundred feet per minute down.

"A man's got to do his best," he said.

"Your best is different from my best. Next time, let me fly the sailplane, OK?"

"No." He meant it, too. He flew every flight we made together, except for a minute or two, now and then. He has made some terrible mistakes, in his time, but there have been some beautiful good flights, too, I have to admit. Mistakes or not, beauty or not, he never lets me fly.

Three hundred feet per minute down, nine hundred feet over the ground.

"This is it," I said. "Get your harness tight as it will go."

He didn't answer, turning toward a paved parking lot in the sunlight. "Maybe not."

The game was over. I knew it. We were dead. Set up for the parking lot, which was too short to land in, and he'll scatter sailplane all over the place. No other place to land . . . wires, trees, roads. Two hundred feet per minute, sliding through seven hundred feet.

"You did it this time, buddy, you really did it!" It was all over but the crash. He wasn't a good enough pilot to land a 1-26 in that space without bending it. A.J. Smith could bring it off, maybe, but this guy, with just a few hours in a

1-26, not a chance. I pulled my harness tighter. Blast, I thought. If I was flying, we would have been safe now in the ridge lift at the hill. But he's flying, with all that romantic bravado, and now we're one minute from disaster.

"Well. How about that," he said. "Lift at last! Two-fifty, three hundred feet per minute up!"

He banked the Schweizer hard to the left, circling in a tight narrow thermal over the parking lot. It was quiet for a long while, as he worked the lift.

"Notice," he said at last, "six hundred feet per minute climb, and we're through twenty-five hundred feet!"

"Yeah. Sometimes you have the most fantastic luck."

"Think it's luck? Maybe so. Maybe not. Believe in lift, never give up the search for it, and I bet you turn out luckier than the man who gives up at a thousand feet. And a fellow hasn't a prayer of reaching his goal unless he somehow learns to find lift for himself, don't you think?"

We rode the lift to four thousand five hundred feet and he set off again on course.

"That little thermal saved your neck," I said, "and you leave it, turn your back on it without so much as a fare-thee-well." I was mostly joking, making a little fun of his dreamer's ways.

"Right. No farewell. Does us no good to stay around after we've gone as high as we can go. Clinging to old lift is for the nonbelievers. Happens over and again. The only real security for a glider is knowing that the sky has other thermals, invisible, waiting. It's just a matter of learning how to find what's already there."

"Hm," I said. It sounded logical enough with four thousand five hundred feet in our bank, but the philosophy was no comfort back there when I thought we had bought us a parking lot.

We pressed along in zero sink for a while, then even that faded and we started down again. We reached the cumulus, all right, but there was no lift there at all. There should have been lift, but there wasn't. I felt hot, suddenly. Below us was the edge of a vast pine forest, hard mountain country—we needed that lift.

"Two-hundred-foot sink," I said. "What do you plan to do now?"

"Guess I'll stay on course. I think that's the right thing to do, sink or not."

The right thing. It's always hard to do the right thing, soaring cross-country. In rising air, for instance, you're supposed to slow down, just when you feel like pushing along on course with the nose down for speed. In falling air, when you feel like holding the nose up, that's when you have it put down, to increase speed and get through the sink as fast as you can. To his credit, then, he set the nose down and penetrated, though we were well out over those hills all spiny with trees, dropping through twenty-five hundred feet with no place to land. He flew as if he had studied textbooks on soaring. Further, he flew as though he trusted that the textbooks were true.

"There's a time," he had told me once, "when you have to believe the people who have already done what you want to do. You have to believe what they tell you, act on it until you're out there proving it for yourself."

I didn't have to ask; that was just what he was doing this moment—believing the diagrams of lift over hills crosswind.

We lost altitude.

"Looks like that cloud might have some lift, off the right wing, couple of miles," I said.

"It might."

There was quiet, for a time.

"Then why don't we cut over there while we still have the altitude to reach it?" I felt like a first-grade teacher with a slow pupil.

"Yes. Well. Look off to the left, too. There's great lift,

ten miles off in that cu. But it's not on course. If we made it over there we could climb, all right, but we'd be ten miles off course, and use all our altitude getting back on. So why make the detour? All we'd do is waste time, go nowhere. That's happened to a lot of good pilots. Won't happen to me, if I can help it."

"Get high and stay high," I quoted at him. He didn't even blink.

What a lousy day! We were down to fifteen hundred feet, in the middle of a bunch of sink, and no place to land but trees. The air was stagnant heavy stuff, like clear granite rock. This was worse than ever. In the parking lot, at least, there would have been people to help us pick up the wreck. Here there wasn't even a lookout tower in the forest. We'd crash unseen.

"What do you know," he said, rolling the glider hard to the right.

"What's up? What are you doing?"

"Look. A sailplane."

It was a pure white 1-26, circling in a thermal not half a mile away. I thought we had been alone, when we left the hill, but there had been somebody out here ahead of us all the time, and now he marked a thermal.

"Thanks, fella, whoever you are." Maybe we both said that.

We slipped in beneath the other Schweizer, and at once the variometer showed two hundred feet per minute climb. It doesn't look like much, on paper, but two hundred feet per minute over a horizon-wide pine forest is a lovely sight. We worked that lift all patient and careful, and by the time we left it, we had had another four thousand feet in the bank. The other sailplane had long since departed on course.

"That was kind of him, to mark that thermal for us," I said.

"What do you mean?" He sounded annoyed. "He didn't mark it for us. He found the thermal for himself and used it for his own climb. You think he made that climb for our sake? He couldn't have helped us an inch unless we were ready to be helped. If we didn't see him, back there, or if we saw him and didn't believe we could use the lift he found, we'd probably be sitting on some pine branch by now."

Just as we left the thermal, we looked down and saw another sailplane gliding down low into the base of it, finding the lift, turning to climb in it.

"See?" he said. "That fellow there is probably thanking us for marking the lift, but we didn't even know he was there till now. Funny, isn't it? We make our own climb, and it turns out that we've done somebody else a favor."

The mountains gave way to flatland toward the end of the day. I was riding along, not thinking much, when he said, "Look there."

There was a wide green field, by the road, and in the center of the field was a sailplane, landed.

"Too bad," he said, with an odd sorrow in his voice.

It startled me to hear him say that.

"Too bad? What do you mean?"

"The poor guy came all this way, and now he's out of the contest, sitting down there in that field."

"You must be tired," I said. "He's not out. The distance he made counts for score, and those points add on to the points he'll make tomorrow and the next day. Anyway, that's not a bad feeling, once in a while, to be down at last and out of the competition for a time, just lying on the grass, resting, knowing you'll fly again."

As we watched, a blue station wagon drove carefully from the road into the field, towing a long narrow sailplane-trailer. It would be a good time. The ground crew would chide the pilot for not doing better, till he lived the flight again for them and proved that he had done his best, every minute. Some things he probably learned, so he might fly a little more skillfully next time. Tomorrow the same pilot would be born again to the competition, at the end of a different towline.

"You're right," he said. "I'm sorry. It's not bad at all. It's exactly right. Forgive me for being so blind."

"That's OK." I couldn't tell if he had been testing me or not. He does that, sometimes.

We tried to stretch our long final glide to the goal, but the sink was worse at dusk and we didn't make it. We touched down in the evening on a lonely pasture, a mile short of goal, but we had done our best and there were no regrets. Even I had no regrets, at the end.

It was still as death when our bright sailplane finally stopped in the green, when the breath of wind about its wings sighed away and was gone.

We opened the canopy, then, I the practical and he the romantic, both in our one pilot-body, stepping free of the body of the sailplane that had taken us through the adventure of the afternoon.

The air tasted light and fresh. We could hear birds in the meadow.

We'd fly again tomorrow, of course, but for the time it was kind of fun to lie down and stretch out in the grass and know that we were alive.

Gift to an airport kid

In my life I had been to four cocktail parties and this one was the fifth and the voice within offered me no mercy. What possible reason, it said, what possible excuse under the name of heaven can there be for you to come to this place? There is one person, clear across the room, who has the mistiest idea of flight, you have one friend in a roomful of strangers intent on paper-thin discussions of national economy and policy and society. You are a long way out of an aviator's element.

A man stood by the mantel this moment, tailored in a double-breasted blazer with polished gold buttons, and talked of a motion picture.

"I liked *Trash,*" he said in a cultured tone, and described in detail a scene that would bore a toad to stone.

What was I doing here? Not fifty feet away, just on the other side of the wall, was the wind and the night and the stars, yet there I stood, soaked in light from electric bulbs and pretending to listen to this man talk.

How can you stand this, I asked me. You are a fake. Your face is turned toward him but you think he's duller than rock and if you had a shred of honesty you'd ask where he finds a point for living if he has to find his values in *Trash* and you should just quietly leave this room and leave this house and get as far away from cocktail parties as you can get and learn your lesson at last and never appear at one of these things again. These affairs are fine for some people but they are not, *not* for you.

Then the crowd kind of swirled, as it does every once in

a while, and I was isolated with a woman crushed with worry over her son.

"He's only fifteen," she said. "He's failing high school and he's smoking marijuana and he doesn't care about living at all. He's blaming me. He'll be dead in a year, I know it. I can't talk to him, he threatens to disappear. He just doesn't care . . ."

It was the first sound of emotion that I had heard all evening, the first hint that anybody in the room was a living human being. Saying what she was saying, casting a line to a scarce-met stranger for help, the woman rescued me from a sea of boredom. I flickered back to when I was fifteen, was eighteen, thinking the world a cold lonely place with no room for newcomers. But about that time I discovered flight, which for me was challenge, was I dare you to survive alone in the sky, and I offer you inner confident quiet if you're good enough to do it, and if you do you'll have a way to find who you are and never be lonely again.

"Has your boy ever flown an airplane, by any chance?"

"No. Of course not. He's only fifteen."

"If he's going to be dead in a year, he sounds like a pretty old man."

"I've done everything I can think of. Racked my brain to get across, to talk to Bill . . ."

I kept thinking of me, age eighteen, changing my life with a two-seater lightplane, with the sound of a small engine at seven a.m., dew in the grass, thin blue smoke from suburban chimneys going straight up into air calm and clear as autumn sky.

"Look. I tell you . . . I have an airplane at the airport, I'll not be leaving till tomorrow afternoon. Mention that to Bill, why don't you? If he's interested, I'll fly him in the Cub, he can get the feel of the controls, how it all works. Maybe he won't like it, but then maybe he will. And if he does, we could go from there. Why don't you tell him there's a flight waiting for him, if he wants it?"

We talked a while longer and there was some faint hope in the woman's voice, clutching at twigs to save her son. Then the evening was over.

I thought about the boy, that night. About the way that those of us who fly have our debts to pay. There's no direct repaying our first flight instructor, for giving a new direction to our lives. We can only pay that debt by passing the gift along, that we were given; by setting it in the hands of one searching as we searched for our place and our freedom.

If he likes it, I thought, the kid can wash and polish the Cub in trade for flying lessons. He can work his way as kids have worked since first there were planes to wash. And one day he'll be set free in the sky and part of my debt will be paid.

I was at the airport early next morning, looking forward to the flight. Who knows? Perhaps he'll be one of those rare natural pilots who will see the idea of flight all in one intuitive flash and know that there is something here that a whole life-style can be built upon. In an hour he'll be flying straight and level, climbs and glides and turns, he can be following through the landing . . .

I thought about this, untying the Cub, making the preflight check, warming the engine. Of course, he might not like it at all. Somehow there are people in the world who do not find that an airplane is a lovely enchanted being, who haven't the faintest wish to be alone in a king's-blue sky and look down upon the countryside. Maybe the kid is one of those. But at least I will have offered my gift, and he would know at least that it is not flight that he is searching for. Either way, it will be some help for the poor guy.

I waited all day. He didn't show up. He didn't even stop by to look at the plane. I'd never know whether he was a natural pilot or not.

"What a thing!" I said later to my navigator, flying cross-country home. "I mean, fantastic! Somebody comes along out of the sky and drops down and offers a free taste of flight, an adventure unlike anything he's done before,

and the kid doesn't even try it! Why, if it were me, I'd have been there at sunrise, pacing back and forth all nervous, waiting!"

It was quiet for a while over a checkpoint, and then the navigator replied, "Did you ever stop to think how he got the offer?"

"What difference does that make how he got the offer? It's the adventure that matters, not how he finds out about it."

"His mother told him about it. His *mother!* Do you think that any fifteen-year-old rebel would ever investigate anything that his *mother* told him about?"

There was no need to answer. What is true has a way of making its point even over the roar of engine and wind.

This is the end of the story. Perhaps by now the kid has found his way or perhaps he's hooked on heroin or perhaps he's dead. The fellow had his own life to live and he lived it the way he wanted to. We can offer a gift, but can never make anybody accept it who doesn't want it.

I'm not discouraged. I'll try again, and maybe someday I can get started on repaying my debt to old Bob Keech, my first instructor, who walked out to meet me one morning at the airport and changed my life with a smile and the words, "Now this is what we call a '*wing*' . . ."

The dream fly-in

It was the strangest fly-in I had ever seen. Maybe it was a dream, it was so strange. There was this new-satin sky, not quite real, with silk-fluff clouds sewn way high (not enough of them to block the sun, which was all lemon light), with green-velvet grass for landing on and concrete white and hard as ivory for taking off. Some big trees around, leafy wide umbrellas for people to sit under and watch the flying. Sandwiches. Cold orangeade.

Parked here and there about this gentle-sloping lawn were airplanes, twenty of them or so, some taxied under the tree shade. Two-place high-wingers, most of them.

I was sitting in this place under the wing of my Cub taking in this strange sight, watching a Cessna flare to land, when this fellow stopped by. He watched the Cessna, too, and then he said, "She's a pretty Cub, you have here. You going to fly in the Test?"

Like anyone who considers himself among the world's most skillful airmen, I'm always game for competition, which is what I figured the Test would be, though I had never heard the word used this way before.

"Of course," I said.

"Glad to have you," he said, and jotted the number of my airplane on his clipboard. He didn't ask my name.

"Is that a sixty-five-horse engine?" he said.

"Eighty-five."

"Height of propeller?"

Now that was a strange question. "Height of prop? Why do you need . . .? Seven feet, I guess."

He shook his head and reached for a tape measure. "What do you do with pilots who come for a Test and don't even know their height-of-propeller?" He walked to the nose of the Cub. "Do you mind?"

"Not at all. I'd like to know."

The tape sighed out, carefully, stretched from the ground to the tip of the propeller. "Nine feet four and a quarter," he said, jotting that number on his board. "And now we need your factor."

"Factor?"

"Performance factor. Wing loading to power loading. Say, is this the first Test you've flown?" He seemed surprised.

"Well, with height-of-prop and factors, I've got to admit, yes."

"Oh! I'm sorry! Welcome aboard! Glad to have you here." He thumbed a sheaf of listing. "Let's see. A Reed Clip-Wing Cub, eighty-five horse . . . here we go. Wing loading eight point five, power loading fourteen point three, and your factor is one point seven." He noted this on his pad. "You don't worry about that," he said, and smiled. "You just fly."

"The Wedge is first. Start engines on the hour. Fly your best." He handed me a thin booklet and then he was gone with his clipboard to an all-white Taylorcraft parked by blanket and picnic basket on the slope opposite.

The booklet was engraved printing in dark blue ink, elegant as an invitation to dine.

THE TEST
PILOTS
OCTOBER 14, 1972

I was skeptical. I don't like my fly-ins quite so organized.

"For those who do not like their fly-ins quite so organized," it said, part way down the first page, "there is a list on page nineteen of conventional fly-ins in the local area. This meeting is designed for those aviators who believe that they are among the best airmen in the world. This meet is the Test to see if they are."

There was a note about the history of the affair, some technical information about the performance factor and judging, and then began an outline of a series of weird challenges, the like of which I had never imagined before.

The booklet pointed out that most pilots don't become good at airplane control without practice, but practice or not, the only way to score well in the test was to fly very skillfully indeed.

I swallowed, at this. I enjoyed thinking of myself among the best, but naturally there are good reasons why it is impossible for me to practice much precision flying. A man has to earn a living, after all.

There was one final comment at the end of the introduction that was perhaps supposed to be funny: "Excuses for poor flying will be heard sympathetically, but will not affect the Test results." I swallowed again, and turned the pages.

THE WEDGE

> TEST: *Altitude control. The Wedge is a tunnel of obstacle ribbons arranged in series across the runway centerline. The highest ribbon is fifteen feet, the others descending in three-inch steps every ten feet down the centerline to form a wedge-shaped tunnel two hundred forty feet long, the lowest ribbon at a height equal to the contestant's propeller height plus two inches . . .*

It went on in detail, about how contestants were disqualified if wheels touched runway, if they strayed from centerline; how there would be no go-arounds allowed and no second attempts. Any pilot who broke more than four ribbons was expected to provide a keg of iced orangeade for the ribbon-setting crew. This was added parenthetically, a kind of traditional joke, but there was no note about the cost of a keg of orangeade.

My brow went suddenly moist, picturing that trap of ribbons coming up at me; and then, remembering that the Wedge was just for openers, which the pilots flew for warm-up, in fun, my brow went cold as death. Propeller height plus two inches.

I thumbed quickly through the booklet, and because self-esteem depends a lot upon one's ability to fly his airplane, I turned from fire to ice to fire again, worse than ever.

The only event I had seen before was a low-speed race, that had happened at Len von Clemm's annual marvel, the antique fly-in at Watsonville, California. In this one, the pilot who took the most time to fly between two marks along the centerline, time corrected by his performance factor, won the points. Not only did one have to know slow-flight, he had to know slow-flight *in ground effect,* even to score.

If this was challenging, though, the remainder of the Test was devastating.

There was to be a Slalom, to find the fastest pilot through a wild-curving one-mile course, the turns marked by giant balloons on strings.

The short-field takeoff lane ended in an angled plywood ramp, six inches high. The pilot chose his own minimum distance to the ramp, made his takeoff with his tailwheel on the ground (or nosewheel lifted, the booklet said . . . there were six nosewheel airplanes in the competition), and catapulted off the ramp. If his airplane flew before the ramp, or if his wheels touched the ground again beyond it, he wouldn't even place in the event.

There was a test in dead-stick spot landings, propeller observed to be at full stop from an altitude of one thousand feet over the field until touchdown beyond a four-foot fence-ribbon.

Next was another dead-sticker: Each plane took aboard only ten minutes volume of his engine's normal-cruise fuel. Each was timed off the ground, the winner to be whichever stayed aloft the longest.

Then came a race through a balloon-and-ribbon obstacle course, where a man had to bank between turn points narrower than his wingspan, leap over red ribbons, dive under blue ones; at least three times were hard-climbing left swerves instantly followed by hard-diving right ones.

It went on and on. There were aerobatic events, formation tests for team flying, even an event in high-speed taxiing. Nowhere was there room for anybody who did not know his airplane, nowhere was there room for a man who talked his ability but could not fly it. For the briefest instant I had the irrational thought that I might very well be one of the latter, but at that moment a green flare was fired aloft and a judge spoke calmly into the loud-speaker, "Start your engines now for the Wedge, if you please."

The white Taylorcraft opposite came to life, the pilot waving cheerfully to an attractive young lady who stayed with the blanket under the tree. He didn't look at all frightened by the tunnel of ribbons. He didn't have to look frightened. He took off in that little airplane, turned around once, like a swimmer turning easily at the end of his lane, and he dived into that Wedge in one smooth motion. For a few seconds he was surrounded and flickered by the ribbons and then he was free, ribbons whipping and fluttering from the wake of his flight but not one of them broken. My throat was very dry.

An Ercoupe took off, turned, and lazily came back into the Wedge to do exactly the same thing. The ribbons weren't even frayed.

I got the Cub engine going, convinced now, as a Cessna 140 shrugged through the tunnel, that it must be a great deal easier than it looked. I have, after all, been flying airplanes for a good many years . . .

Orangeade, by the keg, iced, goes for the staggering sum of $21.75. And ribbons don't jam evenly into an engine cowl. They are cut by the cylinder fins into little shredded smashed pieces that you have to tweeze out with your fingers.

While I completed this job, I decided that the way to practice for the Wedge would be to set up one ribbon, all alone, on my home field. Fly under that till I was good at it, then lower it a bit. The other ribbons were just sham to test a pilot's calm. If one forgets about everything except flying under the lowest ribbon, all the others take care of themselves. But the terror of flying head-on into all that silk (bounced there by a gust, I must have been, though no one else had the same bad luck) is real terror. I think I ducked, as I hit them, and cried out.

The fly-in went right along as though it was not uncommon to have somebody like me entered in the Test. The point, after all, was to prove which pilots are good and which are not quite so good. Any other comments were beside the point, though possibly amusing to spectators.

With just a few exceptions (the low-level race around a ten-mile course, for one), the events of the Test were all held in sight of the airfield . . . people there with *The Test . . . Spectators* booklets to check how the aces and the laughingstocks were faring.

There was not a great feeling of hurry through the fly-in, it was almost languid, with time between events for the pilots to talk, sandwich and potato chip in hand, of the test just past and the one coming up.

My reward was the old maxim of competition, that he who knows least, learns most. Never is it less than delight to stand quiet at the sidelines and listen to a man talk who has just proved, in flight, that he knows what he is talking about.

The pilot of the Ercoupe, for instance. That little airplane, so maligned, in his hands turned and flew like a gazelle in a wide spring meadow.

"There's not all that much to it," he said when I asked. "It's a good airplane. You take a while and get to know it, start to care about it, and you'll find she can do a trick or two, if you let her."

It was the Ercoupe that won the Turnaround . . . flying closest head-on to a crepe-paper wall, then leaping, pivoting in its wingspan, and flying out the opposite direction. I would have bet it was impossible for an Ercoupe to do that.

For all the hard flying, there were no trophies given at the end, no proclamations made about winners. What seemed to matter most to the pilots was how well they flew compared to how well they wished to fly. The reward was not a trophy but a certain knowing that each seemed to value highly.

A sealed envelope was handed to each, which he tucked absently in his pocket to open after the fly-in, if he opened it at all, listing how his flying compared to the others. I, for instance, did not find it necessary to open my envelope.

Do not expect me to go into detail about my performance at the Test, because you see this story is not about my ability as a pilot but about this strange fly-in with all these strange competitions and all these other pilots who had somehow grown to be very good indeed, with their airplanes.

I'm not sure, in fact, that the whole thing wasn't a dream, after all, a remarkably vivid dream. I certainly would have flown much better in all the events had they actually happened instead of being some kind of self-destructive Freudian dream brought on, perhaps, by a slight bounce in an otherwise perfect landing with the Cub.

That must be it. None of this could possibly have occurred. There is no such place as an airfield where lawns slope to runways and one can taxi into tree shade, no such sky as that was, or grass. But most of all there could be no such pilot as the man who flew that Taylorcraft, or the one in the Cessna 140 or that easy gray-haired chap who flew the Ercoupe to the Turn-around and around the Slalom with nothing more than a ruffle of silk ribbon.

I am not so bad a pilot, myself, to have hit any ribbons, anyway. Let me tell you about the time I was flying this Skyhawk. It is another story, not like this silly dream that means nothing because nothing like it ever happened anywhere, but if we ever meet and you want to have a much more honest picture of how good a pilot I in fact am, ask me about the time in the Skyhawk when the engine stopped at ten thousand feet and the only place to land was this *little tiny* field in the trees. And frightened? I wasn't frightened at all because I knew my airplane and it would be child's play even with oil all over the windshield . . .

Ask me sometime about that day in the Skyhawk. I'll be glad to tell you.

Egyptians are one day going to fly

They could have done it, the Carthaginians. Or the Etruscans, or the Egyptians. Four thousand years ago, five thousand years ago, they could have flown.

If you and I were living then, knowing what we know, we could have built an all-wood airframe—cedar, bamboo for spars and ribs, fastened together with dowel pins, glued with casein glue, lashed with thongs, covered with paper or light fabric, painted with root starch. Braided cords for control cables, wood-and-leather hinges, the whole affair light and wide-winged. We wouldn't have needed any metal at all, not even wire, and we could get along as well without rubber or plexiglass.

We might have built the first one swiftly, crude but strong, launched on rails down a hillside into the wind, turning at once into the ridge lift to fly for an hour. Cautious forays, maybe, to hunt thermals.

Then we would have gone back to the shop, having proved it possible, and alone or with Pharaoh's skilled technicians we could have advanced from glider to sailplane to fleets of sailplanes. Learning the principles, the men around us would have discovered flight, would have helped the art in their own way, and before too many years we'd be soaring twenty thousand feet high, flying two hundred miles cross-country, and farther.

Meanwhile, for fun, we'd start to work on metals and fuels and engines.

It was possible, all those years ago, it could have been done. But it wasn't. Nobody applied the principles of flight

because nobody understood them and nobody understood them because nobody believed flight was possible for human beings.

But no matter what people believed or didn't believe, the principles were there. A cambered airfoil in moving air produces lift, whether the air moves today, a thousand years from now, or ten thousand years ago. The principle doesn't care. It knows itself, and is always true.

It's us, it's all mankind that cares, that stands to gain all kinds of freedom from the knowing. Believe that some good thing is possible, find the principle that makes it so, put the principle into practice, and *voilà!* Freedom!

Time means nothing. Time is just the way we measure the gaps between not knowing something and knowing it, or not doing something and doing it. The little Pitts Special biplane, built now in garages and basements around the world, would have been proof of miraculous God-power a century ago. This century there are scores of Pitts Specials in the air, and nobody considers their flight supernatural. (Except those of us to whom a double vertical snap roll followed by an outside square loop to a *lomcevak* have been supernatural right from the start.)

For more of us than care to admit, I'll bet, the ideal of flight lies beyond even a Pitts Special. Some of us just might nourish a secret thought that the very best kind of flying would be to get rid of the airplane altogether, to find a principle, somehow, that would turn us loose all alone in the sky. The skydivers, who have come closest to the secret, also come right straight down, which doesn't quite qualify as flying.

With the mechanical things, the lifting platforms, the rocket belts, the dream is gone—without the tin you're dead, run out of fuel and down you go.

I propose that one day we find a way to fly without airplanes. I propose that right now a principle exists that makes this not only possible, but simple. There are those who say that now and then through history it's already been done. I don't know about that, but I think that the answer lies in somehow harnessing the power that put the whole unseen universe together, that power of which the law of aerodynamics is only an expression in a way that we can see with our eyes, measure with our dials, and touch with the clumsy crude iron of our flying machines.

If the answer to harnessing this power lies beyond machinery, then it must lie within our thought. The researchers in extrasensory perception and telekinesis, as well as those who practice philosophies suggesting man as an unlimited idea of primal power, are on an interesting path. Maybe there are people flying all over laboratories this moment. I refuse to say it's impossible, though for the moment it would look supernatural. In just the same way that our first glider would have looked scary-weird to the Egyptians standing all heavy and small in the valley.

For the time being, while we work on the problem, the old rough fabric-steel substitute called "airplane" will have to stand between us and the air. But sooner or later—I can't help but believe it—all us Egyptians are somehow going to fly.

Paradise is a personal thing

Whether I saw them sauntering out across an acre of concrete to their aircraft, black-leather cubes of flight bags in hand, or silver-flashing at the point of a four-streak contrail way on up at forty thousand feet, I always thought airline pilots the most professional aviators in the world. And "most professional" means highest-paid and that means best. I could never lay a claim to becoming the best pilot alive if I did not fly an airline transport, and besides, the money . . . It is a logically painted portrait into which many a man has walked.

After holding out for years against what I feared would be an exercise in aerial bus driving, boring as sin, I decided that perhaps I was unnaturally prejudiced against airlines. If I am truly excellent in my knowing of flight and sky, I thought, the only proper place for me is on some Boeing flight deck, and the sooner the better. I applied at once to United Air Lines. Gave them all my lists of flying time and certificate numbers and types of aircraft flown, and gave them in full confidence, because I know that if I can do anything at all, I can fly an airplane. I planned to buy the Beech Staggerwing and the Spitfire and the Midget Mustang and the Libelle sailplane all fairly quickly, on an airline captain's pay.

The examinations for the job included one that tested my personality.

Answer yes or no, please: Is there only one true God?

Yes or no: Are details very important?

Yes or no: Should one always tell the truth?

Yes or no. Hm. I puzzled a long time over that test to become an airline pilot. And I failed it.

A United-pilot friend chuckled when I numbly told him what had happened.

"Dick, you take a course for that test! You go down to a school and pay them a hundred dollars and they tell you the answers that the airlines want, and you give the answers that way and you get hired. You didn't answer those questions on your own, did you? 'True or false: Blue is prettier than red?' You answered that by yourself?"

So I planned ways to get around that test. There was not the faintest doubt that I would be a magnificent airline captain, but the test was a tripping stone laid in my way. Just before I paid my money for the answers, though, I idly asked about the life of an airline pilot.

Not a bad life at all. You feel guilty, after a couple of years, taking home a paycheck that size for doing something you consider the best-possible fun. Naturally, you should be a good company man, that's only right. Your shoes are shined and your tie is tied. You follow all regulations, of course, and you join the union, and you keep your hair cut per company policy, and it is not wise to suggest improvements in flying technique to pilots longer employed than yourself.

The list went on, but about that time I began to feel strange little gnawings from within, from the inner man. Why, I could have the greatest attitude in the world for learning the airplane and its systems, I thought, could strive harder than anybody to train uncanny abilities in controlling the machine, could fly it with absolute precision. But if my hair wasn't policy-short, then I wouldn't be quite the perfect man for the job. And if I refused to carry the union card, oddly enough, I wouldn't be a good company man. And if I ever told the captain how to fly . . .

The more I listened, the more I found that United had been right. There was more to it than stick and rudder, instruments and systems. I wouldn't make a good airline pilot, after all, and with a born suspicion of all company policies, I would most likely be a terrible airline pilot.

The airlines had always been a misty sort of Valhalla to me, a land that would always need pilots, that would always yield that diamond paycheck for taking a few hours each month to fly an elegantly-equipped-perfectly-maintained jet transport. And now my little paradise was out the window. They aren't the best, after all. They are company pilots.

So I returned to my little biplane and I changed the oil and started the engine and taxied out to fly, collar unbuttoned, shoes all scuffed, hair two weeks uncut. And up there, perched on the edge of a summer cloud looking out from my cockpit over a peace-green countryside all sparkled with sunlight and washed with limitless cool sky, I had to admit that if I couldn't have an airline pilot's paradise, this one would do till something better came along.

Home on another planet

I had been up in the Clip-Wing, practicing a little sequence: loop to roll to hammerhead to Immelmann, for fun. I was pleased, that day, not to have fallen out of the Immelmann. The trick is full forward stick at the top of the thing, an awkward cross-controlled rudder and aileron for the first half of the rollout, finally reversing rudder to finish. It is not a comfortable figure to fly, but after a while one draws comfort from a good-looking maneuver instead of a pleasant ride. In times past people who have seen my Immelmanns have said, "Gee, you make an awful rollout." I've had to explain that the Air Force never taught any negative-G maneuvers and so I've picked them up by myself and since my learning rate drags without some fanged instructor sitting back there I'm doing good to get the machine right-side-up by the time it's ready to land.

I finished that, a reasonably good sequence with a fair Immelmann, flew around for a while looking out the open side of the cabin at the people down there at work and at school or driving about in tin-shell automobiles along roads barely wide enough to fit. Then landing, and in a moment the engine was as quiet as it had been fifty minutes earlier; a normal end to a normal flight. I got out of the airplane, tied the stick back, the ropes to the lift struts and tail, slid the rudder lock into place.

But then, right in the midst of all that everyday normalcy all at once I had the oddest feeling. The airplane, the sunlight, the grass, the hangars, the distant green trees, the rudder lock in my hands, the ground under my feet . . . they were foreign, strange, alien, distant.

This is not my planet. This is not my home.

It was one of the creepiest moments of my life, that happened for the first time as my hands fell awkward from the rudder lock.

This world seems strange because it is strange. I have only been here for a little while. My deep-secret memories are of other times and other worlds.

What an eerie way to think, I told myself, let's snap out of this, son. But I wouldn't snap. In fact, I remembered mists of this feeling, fragments of it after every flight I've made—the odd thought, returning to the ground, the buried conviction that this planet may be vacation or school or lesson or test, but it is not home.

I have come from another place, and to another place I shall one day return.

It was so absorbing, this odd thing, that I forgot to chock the tires before I left, and so earned a curse from me the next time I went to fly. The vacant numb who forgets to chock his tires, what good can he ever come to?

Yet the ghostly feeling has settled over me time and again since that flight in the Cub. I don't know what to make of it, except that it might be true. And if it is true, if we are all passing through this planet for reasons of experience or learning or tests to pass, what does that mean, anyway?

If it's true, it probably means don't worry. It probably means I can pick up the things I'm so solemn and concerned about in this life and look at them with the eye of a visitor on the planet, and say, these don't really touch me at all. And somehow, for me, that makes a difference.

I didn't think I was the only visitor who has been stopped, rudder lock in hand or rolling through the top of an Immelmann turn, with this kind of harp-jingling shiver that there is a lot more going on than making an aircraft secure or full rudder against the aileron. I knew everyone who flies might have had this knowing, every once in a while, seen strangeness in a world that should by every press of logic be familiar and home.

Right I was. For one day, after a formation flight up over the summer clouds, which was admittedly a handsome sight, a friend said it himself.

"All this talk about going out to space—times like now I get the feeling I'm just coming in. Weird, you know, like I'm a Venusian or something. You know what I mean? That ever happen to you? You ever think that?"

"Maybe. Sometimes. Yeah, I've thought that." So I'm not crazy, I thought. I'm not alone.

It happens more and more often to me now, and I must admit that it is not unpleasant to have roots in another time.

I wonder what the flying's like at home.

Adventures aboard a flying summerhouse

He was selling his airplane to me because he needed the money, but still there were three years of his life in the thing and he liked it and he wanted to hope that I might like it too, as if the plane were alive and he wanted it happy in the world. So it was that after he saw I could fly it safely, and after I had handed him a check, and after waiting for as long as he could stand it, Brent Brown turned to me and asked, "Well, what do you think? How do you like her?"

I couldn't answer. I didn't know what to tell him. Had the plane been a Pitts or a Champ or a fiberglass motor glider I could have said, "Great! Wow! What a lovely airplane!" But the plane was a 1947 Republic Seabee, and the beauty in a Seabee is like the beauty way down in the eyes of a woman who is not a covergirl moviestar—before you see her beautiful, you must begin to know who she is.

"I can't tell, Brent. The airplane flies all right, but I'm still way behind it, it's still pretty big and strange."

Even when the weather cleared and I flew away at last from the snows of Logan, Utah, I couldn't honestly tell Brent Brown that I would ever love his airplane.

Now, nearly a hundred flying hours later, having flown the Seabee across winter America, down the coast to Florida and the Bahamas and back into spring, I can begin to answer his question. We've flown together thirteen thousand feet over mountains sharp as broken steel, where her engine failure would have meant some cool discomfort; we've survived some rough-water ocean takeoffs where my slow beginner's ways in seaplanes could have sent us in large pieces to the bottom. Through these hours I've come to find that the Seabee is generally worthy of trust; perhaps she's found the same is true of me. And perhaps, back in Logan, Utah, Brent Brown could call this the beginning of any real love.

Trust comes not without difficulty overcome. The Bee, for instance, is the largest airplane I've ever owned. With extended wings and droop tips its span is nearly fifty feet. The vertical stabilizer is so high that I can't even wash the tail of the plane without a ladder to climb. Its all-up weight is just over a ton and a half . . . I can't push it alone even across the taxiway, and two men together can't lift the tail-wheel clear of the ground.

Take this huge machine to Rock Springs, Wyoming, let's say, take it there and land in a fifty-degree crosswind twenty gusting thirty (thanking God that the rumors about crosswind landings in Seabees aren't true), struggle it to the parking ramp (cursing the devil that the rumors about crosswind taxiings are), freeze it overnight so the oil is tar and the brakes are stone. Then try to get it flying, come dawn, by yourself. It's like coaxing a frozen mammoth to fly. A Cub or a Champ, you don't need help to get it going, but a Seabee sometimes you do.

After hurling my body like a fevered desperate snowflake against the smooth aluminum mountain of the Bee, hurling it twice and again, I was trembling on collapse and hadn't moved it a fraction of an inch. Then out of the wind came Frank Garnick, airport manager, wondering if he could help. We hitched his snowplow to the mammoth, towed it in compound low till the wheels shattered ice and turned, set a preheater in her engine compartment and a charger on her battery. Half an hour and the mammoth was a fawn, engine purring as though Rock Springs was Miami. You can't always do everything alone; a hard lesson eased by a fellow who didn't mind helping.

With a big airplane one learns too about systems, and how they work. Take the landing gear and the flaps. They all move up and down under the calm physics of the hydraulic system, which is so reliable that it requires no mechanical backup or emergency mode. So that if you squeeze the landing gear down with forty strokes or so of the hydraulic handpump on a night landing to runway 22 at Fort Wayne, Indiana, and touch down with the gear not quite locked, you hear this loud sound—ZAM!—and then a moment later comes a screeching crunching roaring sound wild as freight cars slid sideways on rock.

After you shut the engine down in utter disgust, it gets quiet in the cabin, there in the middle of runway 22, and into that quiet comes a voice, from the tower.

"Do you have a problem, Seabee six eight Kilo?"

"Yeah. I have a problem. The gear collapsed out here."

"Roger, six eight Kilo," comes the voice, pleasant as America itself, "contact ground control on one two one point nine."

You listen to that, and you start to laugh.

Sure enough, just as the factory said, a wheels-up landing on concrete only shaves a sixteenth of an inch from the keel of your new Seabee. Fort Wayne Air Service was there to extend the lesson on help with big airplanes. A clevis had broken in the gear system and a mechanic there hunted me a new one.

"What do I owe you for this?"

"Nothin'."

"Free? You're an airplane mechanic and you're giving me a stranger this clevis free?"

He smiled, thinking of a price. "You're parked at our competitor's place. Next time park here."

Then Maury Miller drove me for nothing all the way back across Baer Field, where John Knight at Consolidated Airways helped me run a gear retraction test, also free of charge. It was either something about the Seabee, or about these people, or about that particular sunrise, but Fort Wayne couldn't do enough to help me out.

"Don't think of a Seabee as an airplane that can land on water," Don Kyte had told me years before. "Think of it as a boat that can fly." A boat that can fly, if you don't care if it's not as fast as, say, a cross-country minie-ball. The Bee trues out at around ninety miles per hour at low cruise, one hundred fifteen at high; this and patience will get you anywhere. At low cruise, the seventy-five-gallon tank holds nearly eight hours flying, at high cruise it's just over five.

Flying his boat over Indiana, Ohio, Pennsylvania, the captain has time to look down and notice tens and scores of little towns right on the edge of blue quiet lakes and wide rivers, and in time he thinks of a way to make a Seabee pay for itself.

"A boat that can fly, folks, just three dollars buys you ten full minutes aloft! It's perfectly safe, your government-licensed pilot, Captain Bach, the air ace, thousands of flights without a mishap, former Clipper pilot on the Hong Kong–Honolulu run, himself at the controls!"

Towns, lakes, breathed away below. Sure enough. It could be done.

After twenty hours in the Bee, I began to feel gingerly at home. Every day the airplane seemed a little smaller, a bit more maneuverable, more a controllable creature than a houseboat in the sky, although the latter is the literal truth. The cabin inside is something over nine feet long, and that before opening the door into the hollow tower under the engine, which adds another three or four feet. The seats recline to make a full double bed. The Seabee Hilton, in fact, is the first flying hotel in which I've been able to stretch out full length and sleep soundly all night . . . a point not to miss in a machine built to spend its nights anchored in wilderness lakes.

The Seabee is fitted with three enormous doors, one right, one left, and then one bow door, set four feet forward of the copilot's seat. According to the owner's manual, this door is for "docking and fishing"; it is also an excellent ventilation door for noons in Bahama waters, when otherwise the cabin overheats in direct sun.

If he's landed by a coast of rocks, or just doesn't feel like leaving his ship, the captain can leave the cabin by any door and stretch out in the sun on a towel or on the warm aluminum along the wingspar, writing or thinking or listening to the waves lap down the length of the hull.

With an alcohol stove, he can prepare hot meals on the cabin top or within, on a galley set on the right half of the flight deck.

I had heard many a discouraging word about the Seabee's Franklin engine, which is odd in that it has a special long propeller shaft and in that it is mounted backwards in the airplane, so that the prop is a pusher. In spite of the words, I've had only one brief engine problem. I noticed in cruise that the engine said mmmmmmmmmm on the magneto-fired sparkplugs when it said mmm-m-mmmm-mmmmm-m on the distributor-fired ones. I reached back into the workshop as I flew along, took out the engine trouble-shooting guide, and deduced that the cause had to be distributor points gone a bit tacky. Sure enough. Next landing I removed the points, replaced them with a new set (which also fits a '57 Plymouth) and the engine said mmmmmmmmmm thereafter, on all sets of plugs.

According to the overhaul manual, the Franklin is good for six hundred hours between overhauls. At two hundred fifty since overhaul, mine burns two-thirds quart of oil per hour at normal cruise. This pleases me because there are Franklins in Seabees which throw that much oil on the vertical stabilizer and are still considered normal.

It's said that a Seabee without the wing extensions is occasionally reluctant to fly. The manual admits that the stock Bee, brand new, can take over 13,690 feet to make a high-altitude water takeoff. Not having flown the airplane without long wings I can't comment, save to say that 68K was flown from Bear Lake, Utah, six thousand feet above sea level, all summer long, with full passenger loads. The long wings and the tips do make a difference.

One special pleasure for Seabee owners resides in a small lever over the pilot's head: the reverse-pitch control for the propeller. It was installed because the Bee, unlike pontoon planes, normally approaches a dock head-on, and so has to leave by backing away tail-first. In the hands of a practiced pilot, reverse pitch makes the plane as maneuverable as a large heavy alligator.

One can use reverse on land, too. The captain taxis into a tight space at the fuel pump, fills up, and then with everybody looking and wondering what happens next, he can yawn, back slowly out of his parking place, and be on his way.

This is hard to top, yet the plane has other and even better features. Last month I flew some twenty-five hundred miles in the Seabee, most of it over the Inland Waterway. It was the most confident secure flying I've done anywhere. Should the engine have failed, I had only to glide straight ahead, or to turn slightly to land on the water. Horizon-wide swamps we flew over, that hadn't enough firm ground for a Cub to land, yet they were all one vast international airport for the Bee: cleared to land whenever we wished, on any runway, upwind, downwind, crosswind, no traffic reported. The airplane is not equipped for instrument flying, but under these conditions it is the best instrument airplane possible.

Following the lee shore of Cape Hatteras, the clouds lowered to two hundred feet and visibility to a bit over a mile—weather one would never consider in a land-plane unless he happened to be flying directly above a hundred-mile runway. In the Seabee, I was. I dropped down to fifty feet over the water, kept my thumb on the map, and pressed ahead like next year's Chris-Craft. When the visibility worsened, I dropped half flaps and slowed. When it worsened still, I decided to land, a matter of easing the throttle back and raising the nose slightly. But just before touchdown, ripples flashing below, I saw a line of light that meant higher ceilings ahead. So we air-taxied along the water for another mile and sure enough, things got better. As I am a chicken in weather, this single feature is my favorite of the Seabee's qualities.

The one dangerous aspect of the airplane, and of most amphibious aircraft, is the other face of its ability to land anywhere. I have talked to three pilots who landed Seabees on the water with the wheels down. Two of them had to swim out of the airplane as it sank upside-down, the third merely had to rebuild the nose section of the plane where it was smashed violently by the sea. For this reason I taught myself to say aloud in every traffic pattern, "This is a land landing, therefore the wheels are DOWN," and, "This is a water landing, therefore the wheels are UP, checked UP, left main UP, right main UP, tailwheel UP. Because this is a WATER landing." I like to say the water-landing check twice before touchdown. It's being a little overcautious, but there is something about the picture of thirty-two hundred pounds on top of me, squashing me against a lake bottom, that I don't mind being overcautious. Then too, aside from being the biggest, the Bee is the most expensive plane I've owned. I do not wish to look down from some rowboat, grappling with a hook for nine thousand dollars of my fortune. If it were a normal-priced Seabee, five thousand to seventy-five hundred dollars, maybe I wouldn't mind.

By the time I had logged fifty hours in the airplane, I had learned how to land it. Thirty hours were spent to believe that I could actually be so high in the air at the moment the wheels first touched; the other twenty were required to discover that just because the wheels had touched didn't mean I wasn't flying the airplane as much as ever. The reason for both learnings was the same—the Seabee has such long oleo shock absorbers that the wheels drop below the place one thinks they ought to be; they roll along the ground a few seconds after the plane is actually flying and for a few seconds before it has actually landed.

The warning is that the Seabee is a high-maintenance machine. I haven't noticed this because I enjoy working on airplanes and don't count the difference between necessary maintenance and work not really required. But here is part of a shopping list made shortly after buying the plane:

Anchor and chain
Raft
Grease gun, grease
Silicon cement
Silicon spray
Weatherstripping
ADF
Scissor jack
Hydraulic fluid
Brake hose
Bilge pump
Bicycle
Cork

There's a story for every item there, even for the cork, which is pressed into the end of the engine compartment oil scupper, to keep black oil from spraying out on the white hull.

The propeller needs to be greased every twenty hours or so, as do wheel bearings and landing-gear fittings. All this can be fun, climbing around and servicing an Alumigrip mountain.

Other elements of Bee flying one learns only by experience. It's a delight, for instance, to taxi up from the water to a lovely virgin beach, but one had best be sure he gets above the high-water line and points the airplane back downhill before he allows it to stop rolling. If not, the captain has an hour's shoveling and messing around with jacks and old boards before his Seabee is unearthed and back in the water.

If the wingtip floats are not sealed around the tops with silicon rubber, water pours in during crosswind water-taxiing, when the downwind float is sometimes completely underwater. Mark the trim indicator overhead for takeoff with different loads; the Bee is very much a trim airplane. Once when the trim froze at high altitude, just a little bit nose-up, I had to ease back the power till the plane flew level by itself—I just didn't have the strength to manually override that trim for more than a few minutes at a time.

Somebody once said that anything worthwhile is always a little bit scary. I was a little bit scared and a little bit cautious about the Bee—how do you know what happens to a summerhouse in flight until you go up and fly one? But in time the captain learns to know its strengths and its quirks, begins to discover its secrets.

One secret of the Seabee I found by chance, that I have found on no other airplane. If one happens to be cruising at ninety-five hundred feet at twenty-two rpm with twenty-two hundred inches of manifold pressure, indicating ninety-seven miles per hour with an outside air temperature of minus five degrees Fahrenheit, and if one is alone in the left seat and if one happens to sing *God Rest Ye Merry Gentlemen* or another song in that frequency range, one's single voice becomes *four* . . . one becomes a kind of airborne Willie the Whale. The strange acoustics have something to do with the thin air, no doubt, and the resonance of the engine at the rpm, but the result is of more than passing interest for those captains who choose to sing only when there's no one else to hear. What other aircraft in the world offers all these features and a full quartet as well, en route to your lake-wilderness hideaway?

I give you, dear reader, the Seabee.

Letter from a God-fearing man

I can keep quiet no longer. Somebody has to tell you people who fly airplanes how tired the rest of us get of your constant talk about flying, about how wonderful it is to fly, and won't we come out on Sunday afternoon and take a little flight with you just to see what it is like.

Somebody has to tell you that the answer is no, we won't come out on the Sabbath, or any other day, to go up in one of your dangerous little crates. The answer is no, we do not think that it is all so wonderful to fly. The answer, as far as we're concerned, is that the world would be a far better place if the Wright brothers had junked their crazy gliders and never gone to Kitty Hawk.

A little bit we can take—we'll forgive anybody for being carried away, when he is just beginning something that he thinks is fun. But this constant day-after-day missionary zeal that you have is just carrying it too far. And that's the word: missionary. You seem to think that there is something holy about knocking around through the air, but none of you knows how childish it all looks to the rest of us, who have some sense of responsibility left for our families and for our fellow man.

I wouldn't be writing this if the situation was getting better. But it is getting worse and worse. I work in a soap factory, which is a fine secure job, with a good union and retirement benefits. The men I work with used to be good responsible men, but now, out of six of us on the Number Three Vat day crew, five have been taken up with this flying madness. I'm the only normal person left. Paul

Weaver and Jerry Marcus both quit work a week ago, they quit together, to go into some kind of business where they think they will tow advertising signs with airplanes.

I pleaded with them, I argued with them, I showed them the financial facts of life . . . paycheck, seniority, union, retirement . . . but it was like I was talking to walls. They knew that they'd lose money (". . . at first," they said. "Till you go broke," I told them). But they just liked the idea of flying so much that it was worth it to them to turn around and walk right out of the soap factory . . . and they've been here fifteen years!

All I could get out of them in explanation was that they wanted to fly, and a sort of strange look that said I wouldn't understand why.

And I do not understand. We had everything in common, we were the best of friends, until this flying business came along—a "flying club" or something like that swept like the plague through the people at the factory. Paul and Jerry dropped out of the bowling league the same day they joined the "flying club." They haven't been back, and now I don't expect they ever will be.

I took the time yesterday, in the rain, to go out to the miserable little strip of grass they call an airport, and talk to the fellow who runs the "flying club." I wanted him to know that he is breaking up homes and business all over town and that if he has any sense of responsibility he will take a hint and move along. That's where I got the word "missionary," and I don't mean that in any nice way, either. Missionary of the devil, I say, for what he's done.

He was inside a big shed, working on one of the airplanes.

"Maybe you don't know what you're doing," I said, "but since you came to town and started your 'flying club,' you've completely changed the lives of more people than I care to name."

For a minute I guess he didn't see how angry I was, because he said, "I just brought the idea. They saw for themselves what flying is like," almost as if it was a credit that so many lives had been wrecked.

He looked to be about forty, but I'll bet he's older, and he didn't stop his work to talk to me. The plane he was working on was made out of cloth, plain old thin cloth, with paint over it to make it look like metal.

"Mister, are you running a business," I said cuttingly, "or are you running some kind of a church here? You've got people running around looking forward to Sunday at this place like they have never looked forward to Sunday at church. You've got people talking out loud about 'being close to God' that have never said the word 'God' as long as I've known them and that's all their lives, most of 'em."

At last he seemed to be getting the idea that I wasn't real happy with him, that I thought he'd better be moving on.

"I'll apologize for them, if you'd like," he said. I could hardly hear him. He twisted himself up under the dashboard of that little airplane, and started taking out one of the dials. "Some of the new students get kind of carried away. Takes them a while to learn not to say what they think out loud, sometimes. But they're right, of course. And you are, too. It is a lot like a religion, flying." He untwisted for a minute and ruffled in his toolbox for another screwdriver, with a smaller handle, and he smiled at me, an infuriating confident smile that said plainly that he wasn't going to be moving on just because responsible people ask him to. "I guess that makes me a missionary."

"Now, that's enough," I said. "I've heard just one time too many this flying-brings-me-near-to-God business. Have you ever seen God on his throne, mister? You ever seen angels flying around your tinker-toy airplane?" I put him a question like that to shake him up, to knock the cockiness out of him.

"Nope," he said. "Never seen God-on-a-throne or angels-with-white-wings. Never talked to any pilots who have, either." He was back under the dashboard again. "Someday when you've got the time, my friend, I could tell you why people talk about God when they get to flying airplanes."

He fell into my trap without so much as a by-your-leave. I'd give him enough rope now, just hear him out, and he would hang himself on a limb of ". . . well . . . er's" and vague mumblings that would prove he came no closer to being a preacher of the gospel than he came to working a vat at the soap factory.

"You go right on ahead, Mister Fly-boy," I said. "Right now. I am all ears." I didn't bother to tell him that I had been to every town revival meeting for the last thirty years, or that I knew more about God and the Bible than he would ever learn in a thousand years, with his tinhorn airplanes. I actually felt a little sorry for him, not knowing who he was talking to. But he had brought it on himself with his ridiculous "flying club" business.

"All right," he said, "let's take a minute and define what we're talking about. Instead of saying 'God,' for instance, let's say 'sky.' Now the sky isn't God, but for the people who love to fly the sky can be a symbol for God, and it's not a bad symbol at all, when you think about it.

"When you're an airplane pilot, you're very conscious of the sky. The sky is always up there . . . it can't be buried, moved away, chained down, blown up. The sky just is, whether we admit it or not, whether we look at it or not, whether we love it or whether we hate it. It is; quiet and big and there. If you don't understand it, the sky is a very mysterious thing, isn't it? It's always moving, but it's never gone. It takes no notice of anything unlike itself." He slipped the dial out of its place, but he kept talking, in no real hurry.

"The sky always has been, it always will be. The sky doesn't misunderstand, it doesn't have hurt feelings, it doesn't demand that we do anything in any particular way, at any particular time. So that's not a very bad symbol for God, is it?"

It is like he was talking to himself, disconnecting lines, easing out the dial, all very slowly and carefully.

"That's a pretty poor symbol," I said, "because God demands . . ."

"Now wait," he said, and I thought he was almost laughing at me. "God demands nothing as long as we ask nothing. But as soon as we want to learn about him, then we run into demands, right? Same way with the sky. The sky demands nothing of us until we want to learn about it, until we want to fly. And then there are all kinds of demands on us, and laws that we have to obey.

"Somebody said once that religion is a way of finding what is true, and that's not a bad definition. The pilot's religion is flight . . . flight is his way of finding out about the sky. And he has to obey those laws. I don't know what you call the laws of your religion, but the laws of ours are called 'aerodynamics.' Follow them, work with them, and you fly. If you don't follow them, no amount of words or high-sounding phrases means a thing . . . you'll never get off the ground."

There I had him. "What about faith, Mister Fly-boy? A man has to have faith . . ."

"Forget it. The only thing that matters is following the laws. Oh, you have to have faith enough to give it a try, I guess, but 'faith' isn't the right word. 'Desire' is a better word. You have to want to know the sky enough to try the laws of aerodynamics, to see if they work. But it's following those laws that matters, not whether you believe in them or not.

"There's a law of the sky, for instance, that says if you taxi this airplane through the wind at forty-five miles an hour, with the tail down, at the proper weight, it is going to fly. It is going to lift right up off the ground and start moving up into the sky. There are lots of other laws that go on from there, but that one is a pretty basic law. You don't have to believe in it. You just have to try taking the plane to forty-five miles per hour and then you can see for yourself. You try it enough times, and you can see that it works every time. The laws don't care whether you happen to believe them or not. They just work, every time.

"You get nowhere on faith, but you get everywhere on knowing, on understanding. If you don't understand the law, then sooner or later you're going to break it, and when you break the laws of aerodynamics you leave the sky mighty fast, I tell you."

He came out from under the dashboard and he was smiling, as though he had a particular example in mind. But he didn't tell me what it was.

"Now, breaking the law, to a pilot, would be the same thing that you might call 'sin.' You might even word your definition of sin as 'breaking the Law of God' or something like that. But the best I can understand about your kind of sin is that it is something vaguely nasty that you're not supposed to do for reasons that you don't very well understand. Well, in flying, there's no question about sin. It isn't hazy in any pilot's mind.

"If you break the laws of aerodynamics, if you try to hold a seventeen-degree angle of attack on a wing that stalls at fifteen degrees, you fall away from God at a pretty good clip. If you don't repent, and get in harmony with aerodynamics before long, you'll have some penalties to pay—like a huge bill for airplane repairs—before you'll ever get back in the sky again. In flying, you get your freedom only when you obey the laws of the sky. If you don't feel like obeying them, you are chained to the ground for the rest of your life. And that, for airplane pilots, is what we call 'hell'."

The holes in this man's so-called religion were big enough to drive trucks through. "All you've done," I said, "is take the words from church and replaced them with your words of flying! All you've done . . ."

"Exactly. The symbol of the sky isn't quite perfect, but it is an awful lot easier to understand than most people's interpretation of the Bible. When some pilot spins out of the top of a loop, nobody says it was the will of the sky that it happened. It is nothing mysterious. The guy broke the rules of smooth flight, trying too high an angle of attack for the weight he had on the wings, and down he went. He sinned, you might say, but we don't consider that a nasty thing, we don't stone him for it. It was just kind of a dumb thing that shows he still has something more to learn about the sky.

"When that pilot comes down, he doesn't shake his fist at the sky . . . he's mad at himself, for not following the rules. He doesn't ask favors of the sky, or burn incense to it, he goes back up there and corrects his mistake; he does it right. A little higher airspeed, maybe, starting his loops. His forgiveness, then, comes only after he corrects his mistake. His forgiveness is that he is now in harmony with the sky and his loops are successful and beautiful. And that, to a pilot, is 'heaven' . . . to be in harmony with the sky, to know the laws and obey them."

He picked a new dial from the bench and crawled back into his airplane.

"You can carry it on as far as you want," he said. "Somebody who doesn't know the laws of the sky would say it's a miracle that a big heavy airplane magically rises off the ground, with no ropes or wires lifting it up. But that's a miracle only because they don't know about the sky. The pilot doesn't think it's a miracle.

"And the power pilot, when he sees a sailplane gain altitude without having any engine at all, doesn't say, 'There's a miracle there.' He knows that the sailplane pilot had studied the sky very carefully indeed, and is putting his study into practice.

"You probably wouldn't agree, but we don't worship the sky as if it were something supernatural. We don't think we have to build idols and offer living sacrifices to it. The only thing we think is necessary is that we understand the sky, that we know what the laws are and how they apply to us and how we can better be in harmony with them and so find our freedom. That's where that joy comes in, that makes the new pilots come down and talk about being close to God." He tightened the lines to the new dial, inspecting them closely.

"When a student pilot starts to understand the laws, and sees them work for him just like they do for all the other pilots, then it's all fun to him, and he looks forward to coming to the airport in the same way that maybe preachers wished their congregations looked forward to coming to church . . . to learn something new, and something that brings joy and freedom and release from the chains of the earth. In short, the pilot, studying the sky, is learning and he's happy and every day is his Sunday. Isn't that how any church-goer should feel?"

At last I had him. "Then your 'religion' says that your pilots are not miserable sinners, soon to suffer hell and damnation, fire and brimstone?"

He smiled again, that same infuriating smile that didn't even give me the comfort of thinking he hated me.

"Well, not unless they spin out of a loop . . ."

He was finished with the airplane, and pushed it out from the shed into the sunlight. The clouds were breaking apart.

"I think you're a heathen, do you know that?" I said, with all the venom I could, and I hoped a lightning bolt would strike him dead, and prove just how heathen he was.

"Tell you what," he said. "I have to check the turn needle in this airplane. Why don't you just come along and we'll take one little flight around the field and you can make up your mind whether we're heathens or the sons of God."

I saw his plot at once . . . to push me out when we were up, or else hit an air pocket and kill us both in his hatred for me. "Oh, no you don't. No getting me up in that coffin! I'm on to you, you know. You're a heathen and you'll roast in the fires of hell!"

His answer sounded as if it was to himself, more than it was to me . . . so soft I could barely hear him.

"Not as long as I obey the laws," he said.

He climbed into that little cloth airplane and started the motor. "You're sure you don't want to go up?" he called out.

I didn't dignify him with an answer, and he went flying all by himself.

So listen to me, you flying people, who talk about your "knowing the sky" and your "laws of aerodynamics." If the sky is God, it is mystery and it is wrath and it will strike you down with lightning and affliction and make you suffer for your blasphemy. Come down out of the sky, come back to your senses, and ask us no more to join you on your Sunday afternoons.

Sunday is a time for worship, and don't you forget it.

Chronology

People who fly *1968*
I've never heard the wind *1959*
I shot down the Red Baron, and so what *1970*
Prayers *1971*
Return of a lost pilot *1969*
Words *1970*
Across the country on an oil pressure gage *1964*
There's always the sky *1970*
Steel, aluminum, nuts and bolts *1970*
The girl from a long time ago *1967*
Adrift at Kennedy airport *1970*
Perspective *1969*
The pleasure of their company *1968*
A light in the toolbox *1969*
Anywhere is okay *1971*
Too many dumb pilots *1970*
Think black *1961*
Found at Pharisee *1966*
School for perfection *1968*
South to Toronto *1971*
Cat *1962*
Tower 0400 *1960*
The snowflake and the dinosaur *1969*
MMRRrrrowCHKkrelchkAUM . . . and the party at LaGuardia *1970*

A gospel according to Sam *1971*
A lady from Pecatonica *1969*
There's something the matter with seagulls *1959*
Help I am a prisoner in a state of mind *1970*
Why you need an airplane . . . and how to get it *1970*
Aviation or flying? Take your pick *1967*
Voice in the dark *1960*
Barnstorming today *1968*
A piece of ground *1961*
Let's not practice *1971*
Journey to a perfect place *1969*
Loops, voices and the fear of death *1970*
The thing under the couch *1970*
The $71,000 sleeping bag *1969*
Death in the afternoon—a story of soaring *1971*
Gift to an airport kid *1971*
The dream fly-in *1970*
Egyptians are one day going to fly *1970*
Paradise is a personal thing *1970*
Home on another planet *1969*
Adventures aboard a flying summerhouse *1972*
Letter from a God-fearing man *1968*

other books by
Richard Bach

STRANGER TO THE GROUND
BIPLANE
NOTHING BY CHANCE
JONATHAN LIVINGSTON SEAGULL
ILLUSIONS
THERE'S NO SUCH PLACE AS FAR AWAY
THE BRIDGE ACROSS FOREVER: A LOVE STORY
ONE